Idaho Mountain Wildflowers:
A Photographic Compendium

Third Edition

A. Scott Earle

Jane Lundin

ISBN 13: 978-0-615-58854-4
ISBN 10: 0-615-58854-9

Cataloging-in-Publication data on file
at the Library of Congress.

Created, designed and produced in the United States.
Larkspur Books
2440 N. Bogus Basin Road, Boise, ID 83702
Tel. 208 344 0079
larkspur1@cableone.net

Distributed by Farcountry Press
P.O. BOX 5630
Helena, MT 59604
1-800-821-3874
1-406-422-1252

Printed in Korea

Front cover: Common bitterroot, pink form
Back cover: Common bitterroot, white form
Frontispiece: *Lupinus arbustus* and the Boulder Range

Contents

Dicotyledons

Apiaceae (Parsley Family) 9

Apocynaceae (Dogbane Family) 17

Asteraceae (Aster, or Composite Family) 20

Berberidaceae (Barberry Family) 60

Boraginaceae (Borage or Forget-me-not Family) 61

Brassicaceae (Cabbage or Mustard Family) 69

Cactaceae (Cactus Family) 78

Caprifoliaceae (Honeysuckle Family) 80

Caryophyllaceae (Pink Family) 84

Chenopodiaceae (Goosefoot Family) 89

Cleomaceae (Cleome Family) 90

Cornaceae (Dogwood Family) 91

Crassulaceae (Stonecrop Family) 94

Ericaceae (Heath Family) 95

Fabaceae (Pea Family) 102

Fumariaceae (Fumitory, or Bleeding Heart Family) 112

Gentianaceae (Gentian Family) 114

Geraniaceae (Geranium Family) 118

Grossulariaceae (Currant Family) 119

Hydrangeaceae (Hydrangea Family) 122

Hydrophyllaceae (Waterleaf Family) 123

Lamiaceae (Mint Family) 131

Linaceae (Flax Family) 130

Loasaceae (Blazing Star Family) 131

Malvaceae (Mallow Family) 133

Nymphaeaceae (Water-lily Family) 136

Onagraceae (Evening-primrose Family) 137

Paeoniaceae (Peony Family) 143

Polemoniaceae (Phlox Family) 144

Polygonaceae (Buckwheat Family) 150

Portulacaceae (PurslaneFamily) 155

Primulaceae (Primrose Family) 158

Ranunculaceae (Buttercup Family) 161

Rhamnaceae (Buckthorn Family) 173

Rosaceae (Rose Family) 174

Rubiaceae (Madder Family) 186

Saxifragaceae (Saxifrage Family) 188

Scrophulariaceae (Figwort, or Snapdragon Family) 194

    Orobanchaceae (Broomrape Family) 194

    Phrymaceae (Monkey Flower Family) 201

    Plantaginaceae (Plantain Family) 204

Solanaceae (Nightshade Family) 211

Valerianaceae (Valerian Family) 213

Violaceae (Violet Family) 215

Monocotyledons

Iridaceae (Iris Family) 217

Liliaceae (Lily Family) 219

    Alliaceae (Onion Family) 219

    Agavaceae (Agave Family) 223

    Ruscaceae (Butcher's Broom Family) 224

    Themidaceae (Cluster Lily Family) 225

    Calochortaceae (Mariposa Lily Family) 226

    Liliaceae (Lily Family) 227

    Melanthaceae (Bunchflower Family) 231

Orchidaceae (Orchid Family) 235

Miscellaneous Plants 240

Bibliography 243

Index 244

Introduction

This edition of *Idaho Mountain Wildflowers*, with the addition of new photographs and descriptive text, is large enough to provide a meaningful cross section of Idaho's wildflowers. It includes almost all the ones that hikers and other visitors to our mountains are likely to encounter.

Added material includes more plants found in northern Idaho's panhandle as well as extensive coverage of the aster family, including plants growing at higher elevations. This family now occupies 40 of the book's nearly 250 pages. (In truth, given time and space, the number of plants in this huge family could have been doubled.)

We have not hesitated to add nonnative plants to the list. Idaho is home to a wide variety of nonnative plants that are at home in our high country. Unfortunately, while some of these non-native "exotics" (or "imports") are attractive additions to our flora, there are others so noxious that we would like to send them back to their native lands! Some of these imported plants arrived with early settlers as contaminants in the fodder that accompanied domestic animals. Others were brought to the New World as medicinal or ornamental garden plants. Many of these are now so well established that they might almost be classified as natives. It doesn't seem right not to include pictures and descriptions of thes plants. Hikers and others who enjoy our mountain flora can't be expected to know "good" from "bad" wildflowers. They want an answer to "what's that plant?" The answers allow them to place the plant where it belongs in the spectrum of western wildflowers. This book should provide those answers.

Acknowledgments

Many individuals who helped with this wildflower project have been recognized before. We are pleased to acknowledge those who also helped with this one:

Botanist Mike Mancuso has over the years and on many field trips shown us a fantastic group of Idaho plants. Photographs of many of them are included in this book. In addition he was kind enough to read our final manuscript. We were pleased to include his many suggestions. Thanks, Mike, for your help; it has been invaluable!

E-mail buddy Sharon Huff kindly allowed us to use her images of the sego lily, *Calochortus nuttallii,* and a photograph, better than the one we used in the last edition, of the crazyweed *Oxytropis besseyi* var. *salmonensis.*

Daughter Wendy and granddaughter Jessica Earle took us to where unique panhandle plants grow so that we could photograph them at the height of their growing season. I thank them for their help.

I am deeply grateful to Kathy Springmeyer, Linda Netschert, Tony Quirini, Ann Seifert and Kelli Street in getting this book to turn out as well as it has. Many thanks to them all!

It's a pleasure to acknowledge John Lundin's help. He drove us to where many of the wildflowers shown in this book were photographed. He never showed anything less than patience in what must at times have seemed like endless photo-botanizing.

Jane Lundin has been an important part of this wildflower project since the beginning. In addition to sharing her photographs she provided suggestions, proof-read the books, and provided the index to this addition. It is my pleasure to add her name to my own as a co-author of *Idaho Mountain Wildflowers.*

And I must again include Barbara Earle. Without her support and encouragement, there would be no book. Love to her and thanks from a lucky man!

A. Scott Earle
March 2012

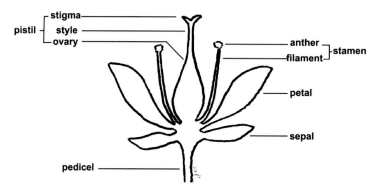

Typical (non-composite) flower

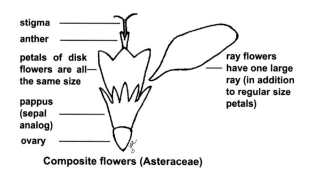

Composite flowers (Asteraceae)

Parsley Family (Apiaceae)

The scientific names Umbelliferae and Apiaceae are both acceptable, although the latter is preferred. There are about 440 genera and 3,590 species in the family. Most are non-woody with thick, often hollow stems. Leaves characteristically wrap about the main stem, as in celery or fennel plants. Small flowers are arranged in flat-topped clusters; the resulting inflorescence resembles the ribs of an inside-out umbrella, the origin of the older family name, Umbelliferae. In some species the flowerheads are compound (i.e., each stemlet divides further). The flowers themselves are almost always radially symmetrical with five small sepals and five petals, although it may take a magnifying glass to make out the details. Some genera contain many similar species, so identification can be difficult, especially as classification is often based on the appearance of the fruit—a technical feature not of much help to amateur botanists. Some species are ornamentals, but the family is valued mostly for its edible plants: carrots, celery, fennel, chervil, parsley, parsnip, etc., as well as herbs used for flavoring including coriander, cumin, caraway, dill and angelica. Despite the many members that have food value, it is unwise to eat any wild umbellifers unless their identification is certain—several are extremely poisonous. The preferred family name, Apiaceae, was derived from *apium*, the Latin word for "parsley."

Lyall's angelica
Angelica arguta Nutt.

Lyall's angelica is a mountain plant that grows nearly to treeline. It blooms from midsummer on, along mountain streams and in nearby wetlands. The name "angelica" was derived from supposed medicinal properties disclosed by an angel. Our species name, *arguta*, means "sharp-toothed" for the shape of the compound pinnate leaves. The leaf-bearing stems angle outward a bit further at each node, where the leaflets come off, an aid to identifying angelicas. The European plant, *Angelica archangelica,* is said to have medicinal value and it is sometimes candied. Use our angelicas and other white-flowered umbellifers at your peril, for they can be hard to distinguish from the poisonous water hemlock, shown on page 13.

David Lyall (1817–1895), whose name is associated with this plant (formerly it was known as *Angelica lyallii*), was an assistant surgeon and botanist on Captain James Ross's voyage of exploration (1839–1844) to the northern Pacific and arctic regions. Lewis and Clark also collected an angelica (probably this species) along the Lolo Trail in northern Idaho on their outbound journey in 1805 and again on their return journey in 1806. *Angelica arguta* grows only in the Northwest, south to California and Nevada, and in the Canadian provinces of Alberta and British Columbia.

Waxy spring-parsley
Cymopterus glaucus
Nutt.

The waxy spring-parsley blooms during snowmelt. The low plants hug rocky ground that holds the sun's heat—a favorable microclimate for cold times. The plant grows only in the mountains of Idaho and Montana, where it is at home at least as high as treeline. The term *glaucus* is Latin for the waxy bloom seen on some plants, as on the skin of grapes, and on this plant's leaves.

Snowline spring-parsley
Cymopterus nivalis
S. Watson

The snowline spring-parsley is a true alpine plant found from treeline to high on the mountain tundra of the alpine zone. It is native to the mountains of Idaho, Montana and Wyoming, and, rarely, to those of Nevada and Oregon. The name *Cymopterus* was derived from two Greek words, *kyma* and *pteron*, meaning "wave-winged" for the plant's winged fruit. The plant's frilly (bipinnate) leaves are a generic characteristic.

Jane Lundin

Douglass's spring-parsley
Cymopterus douglassii
Hartmann & Constance

Another yellow-flowered alpine cymopterus, Douglass's spring-parsley is an uncommon plant that occurs only at high elevations in the Lost River and Lemhi Ranges. Its species name, with a redundant "s," is spelled correctly. The name honors Douglass Henderson of the University of Idaho. The plant may be distinguished from the yellow-flowered *Cymopterus glaucus* (page 10) by its pinnate (feather-like) leaves in which the leaflets do not divide as they do in both of the other *Cymopterus* plants shown on the previous page.

Fern-leaf desert-parsley
Lomatium dissectum
(Nutt.) Mathias & Constance

The fern-leaf desert-parsley (or biscuit-root) is a common species of lomatium. There are several varieties. These differ in appearance, but their identical, striped, pumpkin-seed-shaped fruit allows them to be placed in this species. Their divided leaves (*dissectum* means "divided into many lobes") and their unusually large size—they may be two feet or more tall—will identify the plants as *Lomatium dissectum*; two varieties are shown here. The yellow-flowered plant is *Lomatium dissectum* var. *multifidum* (Nutt.) Mathias & Constance. The varietal name means "much divided." The plant in the inset is *Lomatium dissectum* var. *dissectum*.

Lewis and Clark collected *Lomatium dissectum* on June 10, 1806, near today's Kamiah, Idaho. It can be seen today, lining the walls of the Clearwater Canyon every spring.

Nine-leaf biscuit-root
Lomatium triternatum (Pursh) J. M. Coult. & Rose

The nine-leaf lomatium is easily identified by its leaves, as each leaf divides into three narrow leaflets. These, in turn, divide into three more grass-like leaflets (*triternatum*, from the Latin, means "three times three"). Several varieties are recognized, based on minor differences. Ours is var. *triternatum*. It, like the other lomatiums shown here, blooms early in the spring on gravelly slopes and meadows still moist from snowmelt. This plant occurs in Idaho and every contiguous state as well as in British Columbia, Alberta and California. Lewis and Clark collected all three of the lomatiums shown on these pages and several others as well, during the expedition's return journey in the spring of 1806. None had previously been described, hardly surprising since lomatiums are not found east of the Mississippi River.

Frederick Pursh (1775–1820), the botanist who classified the plants returned by Lewis and Clark, described this plant from a specimen gathered along the banks of Idaho's Clearwater River on May 6, 1806. The generic name, *Lomatium,* is derived from a Greek word that means "fringed," a reference to the appearance of the fruit of some of the species.

Bare-stemmed biscuit-root
Lomatium nudicaule
(Pursh) J. M. Coult. & Rose

The bare-stemmed lomatium (pestle parsnip is another common name) grows on gravelly slopes as high as treeline. Lewis and Clark collected this plant on April 15, 1806, in the vicinity of The Dalles in today's Oregon. Its species name, *nudicaule*, means, appropriately, "bare-stemmed." Native Americans reputedly used this plant to treat consumption, and "Indian consumption root" has been suggested as a standardized common name. This plant is native to Idaho, the coastal states from California to the province of British Columbia, Utah and Nevada.

Western water hemlock
Cicuta douglasii **(DC.) J. M. Coult. & Rose**

The western water hemlock is, emphatically, not a food plant. The related European hemlock (*Conium maculatum* L., an imported species that now grows throughout the United States) was the plant that poisoned Socrates. Ours would have done just as well, for it is extremely poisonous. It grows in most of the United States and in Canada.

It is a handsome plant with dark green, shiny, three-parted, serrated leaves. Clusters of muddy-white flowers resemble exploding fireworks. The plant grows in Idaho's mountains along streams, and in moist meadows as high as treeline. The stems are hollow and are perfect for making whistles—not a good idea, for the poison reportedly has killed children who did so. It has also been the murder plant of choice in more than one detective novel. The Latin word *Cicuta* originally referred to a now unidentifiable poisonous member of the Parsley family—possibly the European hemlock mentioned above. This species name commemorates David Douglas (1798–1834), he of the fir tree, *Pseudotsuga menziesii*, who is said to have introduced more North American plants into English gardens than any other plant hunter.

Cow parsnip, *Heracleum maximum* Bartr.

The cow parsnip (formerly *Heracleum lanatum*) is identified by three-parted, coarsely toothed leaves that may grow to be a foot wide, and by an inflorescence made up of numerous umbels that may be as large as the leaves. *Heracleum* refers to Hercules, who made use of related plants' supposed medicinal properties. Not only are this plant's leaves unusual for their size (the largest of any American umbellifer), but the flowers also are different from those of other Apiaceae, in that those on the margin of the flowerhead are larger than the others and their petals are sometimes bilobed. The cow parsnip grows along stream lines, usually in the company of alders, as high as the subalpine zone. The plants are said to be edible if the furry stalks are skinned first. Cow parsnips grow throughout North America, except the states of the Deep South and Texas.

Slender-leaved lovage
Ligusticum tenuifolium S. Watson

Slender-leaved lovage, also known as licorice-root, grows only in the Northwest and neighboring Rocky Mountain states. The plants grow to subalpine elevations in our mountains, although they are not particularly common. Umbels of white flowerheads are borne atop tall stems. Pinnate (feather-like) compound leaves with narrow, divided leaflets explain the species name, *tenuifolium* ("slender leaf"). As with several other umbellifers, the roots and seeds have a distinctive odor, as reflected in the name "licorice-root." The word "lovage" is more properly applied to the edible European plant *Levisticum officinale*. The related European Scotch lovage (*Ligusticum scoticum*) was also used in the past for flavoring and as a potherb, reminding one of the close relationship of these plants to other culinary umbellifers such as celery, fennel and others. This plant and related American species, such as the fern-leaf lovage, *Ligusticum filicinum*, were used by Native Americans in much the same way.

Great Basin Indian potato
Orogenia linearifolia S. Watson

The Indian potato grows in the foothills and lower mountains of the Great Basin and elsewhere in the Northwest. *Orogenia* was derived from two Greek words: *oros* meaning "mountain" and *genea* meaning "race." The species name, *linearifolia*, refers to its narrow leaves. It is a small plant; the umbels are only about a quarter of an inch across. It is remarkable, however, for blooms develop when thousands of plants emerge and flower simultaneously, always in soggy places, as soon as the snow has melted in early spring. Common names "Indian potato" and "turkey peas" refer to its edible roots. The round to radish-shaped roots are pea-sized or a bit larger. They are tasty (unusual, for many "edible" plants are not) and may be eaten cooked or raw. Plant-hunter Sereno Watson (1826–1892) discovered the two umbellifers shown on this page while a member of the King Expedition.

Western sweet cicely
Osmorhiza occidentalis (Nutt. ex Torr. & A. Gray) Torr.

Crush this plant's flowers and smell their fragrance—indubitably licorice! Although true licorice is derived from another plant, *Glycyrrhiza glabra,* a member of the pea family, the odor is the same. Our plant is easily identified not only by its distinctive fragrance, but also by deep green, three-part, lanceolate leaves and rather dainty yellow flowers, the latter borne in a lacy umbel. Despite the name "sweet cicely," osmorhizas are not true cicelys (or chervils), although they are in the same family and may also be used for flavoring. The western sweet cicely is common throughout our mountain West, and in the western Canadian provinces of Alberta and British Columbia.

Mountain sweet cicely
Osmorhiza berteroi DC.

Three-part, compound, deep green, toothed leaves and tiny white flowers borne in an umbel characterize this imported and now ubiquitous plant. It is found in almost every state and Canadian province growing from sea level to mid-elevations. The roots and flowers of most osmorhizas have a pronounced licorice-like odor (the generic name, *Osmorhiza,* derived from the Greek, means "fragrant root"). This species, however, is odorless, so on first encounter one would hardly suspect that it belonged to the genus *Osmorhiza.* Formerly this plant was classsified as *Osmorhiza chilensis* (it is also found in Chile and Argentina). Recently the species name was changed to *berteroi* to honor an Italian physician, Carlo Giuseppe Bertero (1789–1831).

Northern yampah
Perideridia montana (Blank.) Dorn

The northern, or common, yampah (false caraway is an alternate name) grows in all the Rocky Mountain states as well as in Nevada, the four states of the Northwest, south to California and east to the Black Hills of South Dakota. As with species of *Angelica* and *Ligusticum*, it blooms in midsummer or later on the banks of mountain streams. It is also most likely the edible plant—"a speceis of fennel"—that Meriwether Lewis (1774–1809) saw being harvested on August 26 near Lemhi Pass on today's Idaho-Montana border (where, interestingly, yampah is no longer found). Evidently he did not collect a specimen, or, if he did, it did not survive the journey, for yampah is not represented in the Lewis and Clark Herbarium in Philadelphia.

Swamp white-heads
Sphenosciadium capitellatum A. Gray

Swamp white-heads are tall plants—three or four feet high—found along streams and in moist meadows, from foothills to mountain valleys, blooming in mid- to late summer. They are easily identified by their woolly flowerheads, each made up of many tiny flowers. The attractive little heads are usually white, but sometimes have a pinkish tinge. The leaves are compound, made up of three or more parts. It is the only species in the genus *Sphenosciadium*, found only in Idaho, Nevada, Oregon and California. So far as we are aware, swamp white-heads have no use as a food plant. The scientific name was derived from the Greek words *sphena* meaning "wedge" and *skiada* for "parasol," referring respectively to its wedge-shaped fruits and the plant's umbels. The species name, *capitellatum,* means "little heads." Alternate common names include "rangers' buttons" and "woolly-headed parsnip."

Dogbane Family (Apocynaceae)

Recent taxonomic revisions have combined the Dogbane family (Apocynaceae) with the Milkweed family (Asclepiadaceae), retaining the former name. There are many similarities between dogbanes and milkweeds. For example, both exude thick milky latex when injured and both have fiber-containing seedpods. The reconstituted Dogbane family is made up of approximately 480 genera and 4,800 species. Most grow in the tropics; many are vines, some are trees.

The family is medically important. The anti-hypertensive drug reserpine, for example, is derived from the *Rauwolfia* tree; the anti-cancer drugs vincristine and vinblastine come from periwinkle plants (*Vinca* spp.); strophanthin, a cardiac stimulant, is extracted from species of *Strophanthus*. Other species are important ornamentals, including species of milkweed (*Asclepias* spp.), oleander (*Nerium* spp.), periwinkle and the frangipani tree (*Plumeria*), to list a few. The family is also economically important, producing timber from trees, fiber for cordage, and latex derivatives for various uses.

Both "milkweed" and "dogbane" are old terms. The OED's citations are from 1598 for milkweed and a year earlier, 1597, for dogbane, citing Gerard's *Herbal* for the latter: "Dogs bane is a deadly and dangerous plant, especially to fower footed beasts." Species of both plants grow in our mountains. Several are shown here.

Narrow-leaf milkweed
Asclepias fascicularis Dcne.

The narrow-leaf milkweed is a western plant, found in Idaho—where it seems to be uncommon—Utah and westward to the Pacific coastal states. The plant may be identified by its narrow, opposite leaves and by two or more crowded umbels of dark, reddish-purple flowers borne on short stems that arise from the axils of upper leaves.* The fruit (not shown) is a long pod. The species name, *fascicularis*, from the Latin, means "clustered," referring to the flowers.

---

*The term "axil" as used in this context refers to the upper angle where a leaf arises from the main stem. Cf. "axilla," the anatomical word for the armpit.

Showy milkweed (left, below left)
***Asclepias speciosa* Torr.**

The showy milkweed's range extends eastward from the Pacific coastal states to the midwestern United States and Canada. The plant grows on moist soil and along slow-moving waterways. Its general appearance, attractive flower clusters, milky sap, and, later, the typical milkweed pods, all make it easy to identify. The plant in flower is attractive and is occasionally used as a garden ornamental. This, as well as other milkweeds, is an obligate food plant for monarch butterfly caterpillars. In the past the fluffy seed fibers were used as a hypoallergenic substitute for pillow stuffing. The genus *Asclepias* was named for the Greek physician and minor deity, Asculapius. The plant's species name, *speciosa*, is from the Latin and means "showy" or "handsome."

Spreading dogbane
Apocynum androsaemifolium L.

The spreading dogbane is a loose vine-like plant that spreads along the ground, usually blooming in midsummer. It is often found in the open shade of evergreen forests from mid-elevations to subalpine slopes. Spreading dogbane is easy to identify by its growth habit, by its bright green, smooth-surfaced, opposite leaves, and by its attractive little pink flowers bearing more or less reflexed (bent back) petals. Pick a leaf and you will see the distinctive thick white sap (latex) common to plants in this family. Spreading dogbane grows throughout North America, except in several southern states. In common with most plants that Linnaeus described (the meaning of "L." following the binomial scientific name above), spreading dogbane also occurs in Europe; most likely it is the plant Gerard termed "dogs bane" (see page 17).

Hemp dogbane, or Indian hemp, *Apocynum cannabinum* (not shown), is a related, similar species that also grows in Idaho (and in every state and most Canadian provinces), preferring disturbed or previously cultivated ground at lower elevations. Native Americans used the fuzzy fiber contained in the plant's seedpod to make cordage, explaining its common names.

Aster, or Composite Family (Asteraceae)

The common names for this family: sunflower, composite and aster family; and the scientific names Compositae, and Asteraceae, are all correct, although the terms "aster family" and "Asteraceae" are preferred today. It is the largest family of flowering plants, made up of 1,530 genera and 23,800 species. All family members have a common characteristic: each bloom or flowerhead (commonly, but incorrectly, referred to as the plant's "flower") is made up of many tiny flowers. Those in the center of the flowerhead are "disk flowers," and those on the edge, each with a single strap-like petal, are "ray flowers." In some species ray flowers occupy the entire flowerhead; conversely, others have only disk flowers. The Asteraceae have other characteristics that can help with identification. These include protruding Y-shaped styles; multiple pointed small leaves, or "bracts" (collectively forming an "involucre") that cup each flowerhead; and often a feathery "pappus" attached to each seed that aids in seed dissemination (as, notably, with dandelions). It is not always easy to recognize members of the aster family for some flowerheads may be tiny, resemble those of other plants, or are in other ways atypical. Many of the Asteraceae are cultivated as garden ornamentals: zinnias, chrysanthemums, daisies, and asters, to name a few. Some are important food plants: lettuce and other greens, artichoke, and the food-oils obtained from sunflowers and safflowers. Still others are troublesome weeds: ragweed, knapweed, thistles, burdock, common dandelion, and many others. It would take a book to describe all of the Asteraceae that grow in Idaho's mountains; the ones pictured here can only serve as a sampling.

Alpine aster*
Oreostemma alpigenum (Torr. & A. Gray) Greene
var. *haydenii* (Porter) G. L. Nesom

The alpine aster is a subalpine to alpine species that grows in clusters in open spaces. It is found in the northwestern states and in Nevada and California. Botanists recognize three varieties; only this one grows in Idaho. The plant has a basal crown of linear gray-green leaves. These, the stem, and purple-tinged bracts are covered with fine hair most noticeable on young plants. The generic name, *Oreostemma,* derived from two Greek words, means "mountain crown." To avoid confusion with other alpine asters, the name "tundra mountain crown" has been suggested for this plant.

---

*Following recent taxonomic revisions, many Asteraceae formerly classified as species of *Aster* are now included in several smaller genera. It may help to think of the species formerly in genus *Aster* as aster-related plants (or maybe "asteroids"?), an informal classification supported by many common names that have "aster" in them. Although there is considerable overlap, this group of plants has the following characteristics that separate them from the rather similar fleabanes (*Erigeron* spp.): the "asteroids" tend to bloom later and they have "shingled" bracts lined up in irregular rows under their flowerheads.

Leafy aster
Symphyotrichum foliaceum (Lindl. ex DC.) G. L. Nesom
var. *apricum* (A. Gray) G. L. Nesom

The leafy aster, a tall plant, blooms throughout the West, from midsummer on. The plants have large, stemmed basal leaves and smaller, clasping (stemless) leaves higher up. Each stem bears one flowerhead, most commonly with fifteen rays. As the flowers mature, their bright yellow disk becomes brownish and the petals darken to a rich, deep purple, so much so as to be a distinguishing feature for the plant. Several varieties are recognized by minor differences in their morphology. Var. *apricum* (*apricum* means "sun-loving") shown here has purple-margined bracts (the small leaflets that cup the flowerhead). It is a common plant along our trails.

White prairie aster
Symphyotrichum falcatum
(Lindl.) G. L. Nesom

The white prairie aster is distinguished by clusters of white flowerheads that tend to be borne on the same side of tall stems. The plants spread by short rhizomes (shallow root-like stems) that also contribute to their clustered appearance. Although commonly found on the plains, the plants also grow in our mountains. This one was photographed just west of Lolo Pass in north-central Idaho. The species name, *falcatum*, means "sickle-shaped." It is not clear why it was applied to this plant.

Thickstem wood-aster
Eurybia integrifolia (Nutt.) Nesom

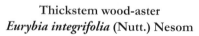

Formerly *Aster integrifolius*, the new generic name from the Greek means "wide-spreading," referring to its form as a loose clump. This species is found in the Northwest, California, Nevada and Colorado. The plants grow to be two feet or more high. Their lower leaves are large and short-stemmed. The stem leaves, however, become increasingly smaller and clasping (i.e., the leaf base grows partway around the stem). The upper leaves and stem tend to be glandular imparting a sticky feeling when touched. This wood-aster is usually found growing in meadows and open woods at mid-elevations.

Western American aster
Symphyotrichum spathulatum (Lindl.) Nesom

Formerly known as *Aster occidentalis,* this plant grows from the montane to alpine zones. It is found in most western states and provinces although it is not particularly common in Idaho. The plant is distinguished by its height; it may grow to be three feet tall or higher. It favors moist mountain meadows or dried seasonal streambeds. The plants are leafy with smaller basal leaves, rounded at their ends (*spathulatum* means "spoon-shaped"), becoming larger and more lanceolate higher on the stem. The flowerheads have twenty to fifty purple rays and a small yellow disk. The plant blooms from midsummer on.

Elegant aster
Eucephalus perelegans
(A. Nelson & J. F. Macbr.) W. A. Weber

Hoary aster
Machaeranthera canescens var. *canescens*
(Pursh) A. Gray

The elegant aster blooms from midsummer on, growing as high as the subalpine zone. Its five (occasionally eight) rays make it quite distinctive. The elegant aster was collected in 1834 and described in 1841 by botanist Thomas Nuttall (1786–1839), who had recently resigned from the faculty of Harvard College to join Boston businessman Nathaniel Wyeth (1802–1856, for whom the *Wyethia* was named) on a journey west to Oregon. The plant grows from Oregon to Montana and south to Nevada, Utah and Colorado.

The name *Machaeranthera* was derived from two Greek words meaning "sword" and "anther," for the plant's sharp-pointed anthers. The hoary aster blooms in summer's heat, from the end of July through August and into September, in dry places. A furry stem (the Latin *canescens* implies "covered with gray hairs"), spiny bracts, variably white-based to fully purple rays and an orange disk help to identify the plant. Meriwether Lewis collected a hoary aster (another variety, var. *incana*) on the Columbia River in October 1805, near today's The Dalles, Oregon.

Alpine aster
Ionactis alpina
(Nutt.) Greene

One of the problems with common names is that they are often duplicated; there are, for example, several "alpine asters." The name "lava-aster" has therefore been suggested (it is commonly seen growing in the lava beds of Craters of the Moon National Monument), although they are not restricted to laval ground. More commonly the plants are found in the company of sagebrush, growing as high as the subalpine zone. Characterized by small, clustered gray-green leaves and thirteen or so rays, it is a native of Idaho, Montana, Wyoming, Utah, Oregon and Nevada. It was previously classified as *Aster scopulorum.*

Rocky Mountain aster, *Ionactis stenomeres* (A. Gray) Greene

The Rocky Mountain aster's attractive flowerheads usually bear thirteen (occasionally more) narrow rays. The word *stenomeres,* from the Greek, means "narrow parts," presumably referring to the rays and to the narrow, same-sized leaves that ascend the stem. *Ionactis* was also derived from the Greek and means "violet rays" (cf. "actinic"). The Rocky Mountain aster grows only in Idaho, Montana, Washington and the province of Alberta.

Fleabanes or daisies, *Erigeron* **species**, differ from the "asteroids" in several ways; most notably they bloom earlier and the bracts that subtend the flower are regular in configuration. The genus name, *Erigeron*, derived from the Greek (*eri+geron*), means early old man—fleabanes bloom early and have grayish, hairy foliage. The botanically preferred common name, fleabane, was derived centuries ago because several daisy-like plants, notably species of *Chrysanthemum* and *Tanacetum*, contain insecticidal pyrethrins.

Long-leaf fleabane
Erigeron corymbosus Nutt

The long-leaf fleabane, while often classified as a subalpine plant, is more commonly seen growing in foothills. It is characterized by wavy, somewhat hairy, linear leaves. The species name, *corymbosus*, is Latin for "cluster" (of fruit or flowers). Its botanical meaning implies a cluster of plant elements that are in a flat plane. The plant grows in all the northwestern states, in Utah and Nevada and in British Columbia.

Rockslide fleabane
Erigeron leiomerus Gray

Although the U.S. Department of Agriculture and others list this plant's common name as rockslide yellow fleabane, that's a bit of a misnomer because its flowerheads, as those shown here, are often white. The rockslide fleabane is a high-altitude daisy that is found growing in rocky places in the mountains of central Idaho and various mountain ranges of all the adjacent states and British Columbia.

Idaho fleabane
Erigeron asperugineus
(D. C. Eaton) A. Gray

The Idaho fleabane is a lovely little subalpine plant that blooms in early summer. The name *asperugineus* means "rough," referring to the plant's brittle-feeling stems and crisped leaves. It occurs in northern Nevada (where it was first collected), but is more common in Idaho, especially in the mountains of the central part of the state, and in western Montana.

Showy daisy
Erigeron speciosus
(Lindl.) DC.

The showy daisy is common in our mountains, where it grows in large groups crowded with many flowerheads. These are about two inches across, with 60 to 150 narrow rays that range in color from purple to nearly white. Alternating lanceolate leaves ascend the stem. Showy daisies are distributed throughout the Rocky Mountains and the Northwest. "Aspen daisy" has been suggested as a standardized common name for the plant. David Douglas introduced the showy daisy into England and it is grown there today as an ornamental. The name *speciosus* means "splendid."

Bear River fleabane
***Erigeron ursinus* D. C. Eaton**

The Bear River fleabane's flowerhead is larger and showier than those of most plants in genus *Erigeron*. It typically has a basal cluster of narrow, club-shaped leaves and linear stem leaves. The stem is stout, bearing a single flowerhead whose involucre (the small leaves that hold the flowerhead) may be somewhat sticky. A broad yellow disk is surrounded by one hundred or so purple rays. The plant grows in the high mountain ranges of eastern Idaho, western Montana, Utah, Colorado and Wyoming. The Bear River, from which both common name and scientific species names were derived (*ursinus* means "bear" in Latin), flows from Utah's Wasatch Range to Bear Lake on the Idaho-Utah border and thence to the Great Salt Lake.

Dwarf mountain (or cutleaf) daisy
Erigeron compositus Pursh

The dwarf mountain (or cutleaf) daisy is a miniature plant that grows on rocky ground and exposed slopes. While it grows at all elevations, it thrives as an alpine plant, growing above treeline on alpine tundra. Typically it forms compact clumps that bloom from mid-spring to midsummer according to the elevation. The plant grows in most western states, through Canada to Alaska, and east to Greenland. Lewis and Clark were the first to collect the cutleaf daisy, possibly in the fall of 1805 (as a dried plant), or in the spring of 1806 while they were camped on the Clearwater River near today's Kamiah, Idaho. Because it is both attractive and hardy, this little plant is used as an "alpine" in rock gardens. The name *compositus* means "compound," referring to its three-parted, or ternate, leaves made up of deeply divided leaflets.

Evermann's fleabane
Erigeron evermannii Rydb.

Evermann's daisy is a good-sized composite with basal lanceolate leaves, a leafless, furry stem and a single flowerhead. It is a true mountain flower, one that is found only at high elevations in Idaho, Montana (where it is rare) and Alberta. Its rays are sometimes tinged blue or pink, although ours are more often white, as shown here.

Barton Warren Evermann (1853–1932) was a naturalist, best known as an ichthyologist. He published an important ichthyological text, *Fishes of North and Middle America,* in 1900.

Coulter's daisy
Erigeron coulteri Porter

Coulter's daisy resembles Evermann's daisy in that a single white-rayed flower-head is borne on each stem. It differs, however, because its stems are not bare but have several to many lanceolate leaves that become clasping (stemless) as they ascend the stem. There are also more ray flowers, usually from forty to one hundred, whereas Evermann's daisy usually has forty or fewer. Although the involucre is not shown in our photograph, its appearance may be helpful in identifying the plant. It consists of narrow, regularly placed, modified leaves that cup the flowerhead. Uniquely, there usually are fuzzy black hairs around the base of the involucre. Coulter's daisy grows at high elevations, often along streams, and is present in all the Rocky Mountain states, as well as in Nevada and California.

John Merle Coulter (1851–1928) for whom the plant was named, started his career as botanist to a geological expedition that explored the Rocky Mountains in 1872–73. He later became an academic, holding chairs in botany at several colleges. He is known today for his comprehensive Manual of Rocky Mountain Botany (1885).

Jane Lundin

Mountain townsendia
Townsendia alpigena Piper

The townsendias are closely related to the fleabanes. The mountain townsendia shown here is a small, alpine rock plant that grows at or above treeline in the mountains of Idaho, Montana, Wyoming, Utah, Oregon and, rarely, in Nevada. The plants are characterized by relatively large flower-heads borne one to a stem. Their central disks are prominent, and the rays range from a light purple color (usually) to near white. Its small, ovoid basal leaves are covered with fine hair, and the involucral bracts below the flowerhead are striped, a distinguishing feature. Twenty-seven species of *Townsendia* are now recognized; none grow east of the Mississippi River. David Townsend (1787–1858), for whom the genus was named, was an amateur Pennsylvania botanist.

Heartleaf arnica
Arnica cordifolia Hook.

While the origin of the generic name *Arnica* is unknown, its species name, *cordifolia,* makes sense, for it means "heart-shaped leaf," and its leaves are unmistakably that shape. It is one of our commonest mountain composites. The plants bloom from late spring until midsummer according to altitude. They prefer the open shade of evergreen forests, but occasionally overflow onto neighboring meadows. Typically the stems lie close to the ground (they are prostrate), turning upward to flower. Eight to thirteen wide, pointed, bright yellow-orange rays are borne on each flowerhead. Heartleaf arnica grows in much of the West, ranging from the Yukon Territory to New Mexico.

Streambank arnica
Arnica lanceolata Nutt.
var. *prima* (Maguire) Strother & S. J. Wolf

Arnicas are not difficult to identify, at least at the generic level. Their flowerheads are borne at the end of one to several stems and are fairly large with eight to fifteen bright yellow rays and a well-defined, rounded (turbinate) disk. Opposing leaves—an identifying feature—are given off at intervals along the stem; in some species the leaves also form a basal cluster. The streambank arnica resembles several similar species. It is tall, thin-stemmed and, as its common name suggests, grows along mountain streams, typically at high elevations. It is easily identified by where it grows, by the stemless leaves that "clasp" the main stem (*amplexicaulis*, an older species name, means "stem-embracing") and by its finely serrated leaf margins.

The sap of arnicas may irritate; tincture of arnica, usually derived from the European plant, *Arnica montana*, has been used for centuries as a rubifacient, an external application useful for treating painful sprains and bruises.

This plant was formerly classified as *Arnica amplexicaulis* Nutt. and is so listed in older guide books.

Twin arnica
Arnica sororia Greene

The twin arnica takes its species name from the Latin word for "sister," presumably because two (or more) flowerheads arise from a common stem. It is a meadow plant, growing to fairly high elevations; it may be identified by this growth habit and by its cluster of large basal leaves and smaller opposed, lanceolate stem leaves. The twin arnica is closely related to a similar, taller and larger-leaved species, shining arnica, *Arnica fulgens* Pursh (not shown), that has much the same range throughout our western mountain states and Canadian provinces. The two are so closely related that until recently the species shown here was considered to be a varietal form of the larger plant.

Slender arnica
Arnica gracilis Rydb.

Arnica gracilis (formerly *Arnica latifolia* var. *gracilis*) is a plant whose habit—cluster-forming, low-growing, nestled among rocks—is typical of many other alpine and subalpine plants. As illustrated, it is at home on rocky ground, which protects the plants and holds the sun's heat. By looking at the plants closely, one can see that small, broadly lanceolate leaves arise opposite each other, typical of arnicas in general. The slender arnica is found in the mountains of all the northwestern states and provinces, and south to Wyoming, Colorado and Utah.

Spear-leaf arnica
Arnica longifolia D. C. Eaton

The spear-leaf arnica (left and below) is a high-altitude, cluster-forming composite. It blooms in mid-August, usually close to water, in or near seep-springs, lakes and slow-moving streams. The plant's preference for sheltering rocks, its proximity to water, and its long, opposing, pointed leaves (responsible for both common and specific names) set it apart from other high-altitude clustered composites. (The shrubby golden-weed, shown on page 38 is the plant it most closely resembles.) The leaves of both plants are sticky, and both give off an odor when crushed. The medicinal odor of the arnica's leaves is not nearly as strong, however, as that of the highly aromatic goldenweed.

The spear-leaf arnica is native to all the central and northern Rocky Mountain states, as well as Nevada, California, Alberta and British Columbia, although it is rare in the Canadian provinces.

Mountain (or false) dandelion
Agoseris glauca (Pursh) Raf.

Even though it lacks a central disk and is somewhat similar in appearance to a dandelion (*Taraxacum officinale*), the mountain dandelion is not closely related to the weed. The mountain dandelion is characterized by a long bare stem, a terminal flowerhead and a few linear leaves that spring from a basal rosette. Two varieties are recognized based on relatively minor differences (width and shape of leaves, plant size, etc.). Ours is var. *glauca*. The plant grows at all elevations, as high as treeline. The terms "mountain dandelion" or simply "agoseris" are used for the plant. Because *agoseris* was derived from two Greek words that mean "goat-chicory," the common name of "pale goat-chicory" has been suggested for the plant. The species is distributed throughout the western states, east to the Great Lakes and north to Yukon Territory.

Orange mountain dandelion
Agoseris aurantiaca (Hook.) Greene

The orange agoseris is a relatively uncommon plant that you'll see from time to time during the summer months, usually at higher elevations. It is an eye-catcher because its burnt-orange hue is unusual among the Asteraceae that grow in our area. We include it here with the yellow composites because of its close relationship to the mountain dandelion, shown above. The species name, *aurantiaca,* from the Latin, means "orange-red" (the word is cognate with "orange"). Its leaves are a bit wider than the agoseris shown above, but otherwise the two are similar. This plant also grows in the western mountain states, north to Alaska and western Canadian provinces, west to the Pacific coast and south to California and New Mexico. The ungainly term "orange-flower goat-chicory" has been suggested as a standardized name.

Arrowleaf balsamroot
Balsamorhiza sagittata
(Pursh) Nutt.

This robust and ubiquitous plant grows in most western states and provinces, from sea level to treeline. Its clustered, arrow-shaped, gray-green leaves and showy flowerheads make it easy to identify. The plants bloom from spring through midsummer on ever higher mountainsides. Its common and scientific names are from its arrow-shaped leaves and the balsam-like odor of its roots. Native Americans used its roots and shoots for food (although it is far from tasty). Lewis and Clark twice gathered specimens of balsamroot in spring and summer 1806.

Largeleaf balsamroot
Balsamorhiza macrophylla Nutt.

Although the arrowleaf is by far the most prevalent species of balsamroot, there are nine other western species. The largeleaf balsamroot (also known as cutleaf balsamroot) is a Great Basin species that crosses into Idaho. It is similar to the arrowleaf balsamroot, differing chiefly in the shape of its large, incised, pinnate leaves. The species name, from the Greek, reflects its common name, "largeleaf."

Hooker's balsamroot
Balsamorhiza hookeri (Hook.) Nutt.
var. *hispidula* (Sharp) Cronquist

Hooker's balsamroot is found in many places in the West. Six varieties are recognized; some are quite localized in distribution. The variety shown here grows in south-central Idaho and south into Utah and Nevada. It is a small plant with pinnate (feather-like) leaves. Its stems and leaves are hairy, as its varietal name *hispidula* ("covered with stiff hairs") suggests.

Scabland fleabane
Erigeron bloomeri A. Gray

The scabland fleabane is usually found growing in barren, rocky sites in the mountains and foothills of Idaho, Nevada, Utah and the Pacific coastal states. As the illustration suggests, it is characterized by narrow basal leaves and naked stems that each bear a single rayless flowerhead. The species name honors an eminent California botanist, Dr. Hiram Green Bloomer (1819–1874), who collected the plant in Nevada, near Virginia City.

Line-leaf daisy
Erigeron linearis
(Hook.) Piper

The line-leaf daisy (also known as the desert yellow fleabane) is, like the cut-leaf daisy shown on page 28, a small plant that grows in discrete clumps. The two are often seen growing together. This plant also prefers exposed gravelly slopes where it may grow in profuse numbers from late spring into the summer, as high as treeline. As with many plants that are adapted to dry places, both leaves and stems feel brittle. The common and scientific species names describe its thin, "linear" leaves. The line-leaf daisy is restricted to the northern Rocky Mountains (including British Columbia) and western coastal states.

Common sunflower
Helianthus annuus L.

The common sunflower, originally native to North America, has now spread throughout the world. The plant is immediately recognized by its large flowerhead, broad hairy leaves and tall stems. *Helianthus* is from the Greek words for "sun" and "flower"; *annuus* is Latin for "annual" (botanical names often mix Greek and Latin). Sunflowers bloom from midsummer on and are found throughout the United States and in Canada north to the Arctic Circle. The plants grow at least as high as the montane zone, in open fields and along fence-lines and roadsides. Sunflowers have long been cultivated for their seeds, and more recently for the oil that the seeds contain. They are considered to be a noxious weed in Iowa.

Nuttall's sunflower
Helianthus nuttallii Torr. & A. Gray

The genus *Helianthus* is a large one, made up of sixty-seven species and many varieties; several grow in Idaho. Nuttall's sunflower is a fairly common, tall, perennial plant characterized by sunflower-like flowerheads and narrow, lanceolate, mostly opposed leaves. The plants prefer moist or recently moist soil and grow to fairly high elevations in our mountains. It is found in all our western states, except several in the south, and in all the lower Canadian provinces, east to Quebec. Two other varieties are recognized and are known by their varietal names in other parts of the country.

Rocky Mountain dwarf sunflower
Helianthella uniflora (Nutt.) Torr. & A. Gray

When Thomas Nuttall published the first description of this plant in 1834, he classified it as a species of sunflower, *Helianthus uniflora*. Later, it was reclassified as its own genus, *Helianthella* (the diminutive of *Helianthus*). Typically the dwarf sunflower has several long stems that arise from a thickened, persistent base (caudex). Large, opposing, lanceolate leaves are given off at intervals along the stems; each stem is topped by a single (*uniflora*) sunflower-like flowerhead. Although this plant is not classified as a high-altitude species, it grows at least as high as the montane zone in the mountain ranges of Idaho and other Rocky Mountain and Pacific coastal states, north to British Columbia and south to California, Arizona and New Mexico.

Curly cup gumweed
Grindelia squarrosa (Pursh) Dunal
var. *quasiperennis* Lunell

The curly cup gumweed is an odd plant because its flower cups contain a viscous, resinous fluid. Native Americans used the resin to treat skin conditions and respiratory problems. The leaves were used for tea, and the buds were chewed as gum. The Latin species name, *squarrosa*, means "bent at right angles," referring to bracts that turn outward at the base of the heads, forming a "curly cup." Gumweed grows along mountain roadsides, often in great numbers, blooming from early summer into the fall. The name *Grindelia* honors David Hieronymus Grindel (1776–1836), a Russian botanist. Gumweed grows throughout the United States, except in a few southern states; it is also found in Canada and Mexico. Lewis and Clark collected another variety of this plant in present-day Nebraska on August 17, 1804.

Indian blanket-flower
Gaillardia aristata Pursh

The Indian blanket-flower, or gaillardia, is native to Idaho and the other Rocky Mountain states, west to California and east across the country in the nothern tier states and Canadian provinces. Meriwether Lewis collected this species in today's Montana, near the Continental Divide on July 7, 1806. The plant is easily identified by its reddish-brown-based, three-lobed rays and its bristly, reddish-brown disk (*aristata* means "bristly"). The name *Gaillardia* honors an eighteenth-century French magistrate, Gaillard de Merentonneau (also spelled Charentonneau), who had an interest in botany. *Gaillardia aristata* is often confused with the colorful firewheel, *Gaillardia pulchella*, a common garden plant that is native to our southern states. It may escape and persist locally in our area for several seasons.

Shrubby goldenweed
***Ericameria suffruticosa* (Nutt.) G. L. Nesom**

The shrubby goldenweed (formerly *Haplopappus suffruticosus*) is a summer-blooming plant that grows as high as treeline, sometimes turning barren, south-facing slopes bright yellow. The flowerheads are few-rayed (five to nine) with bristly central disks. Crisp-edged leaves are covered with fine hair. The plants have a very strong, but not unpleasant aromatic odor that may fill the air even before the plants have bloomed. The shrubby goldenweed grows in the northern Rocky Mountains south to Colorado and west to California. (Certain botanists have suggested that the common name of the goldenweeds in this genus be changed to golden bush to avoid confusion with plants in genus *Stenotis*.)

Jane Lundin

Whitestem goldenweed
***Ericameria discoidea* var. *discoidea* (Nutt.) Nesom**

The whitestem goldenweed (formerly *Haplopappus macronema*) has, in common with other members of the genus, a bristly flowerhead. Unlike the shrubby goldenweed shown above, this fragrant subalpine plant has only disk flowers explaining its species name. The involucre cupping the flowerhead is made up of rather sticky, narrow, leaf-like bracts. These are typically puberulent (covered with short hairs), as are the leaves and stem. There may be a single flowerhead as in the plant shown here. More commonly, however, there are three or more, borne on thick, whitish stems. The plants grow on rocky ground and are found in most of the western states neighboring ours, except for Washington and, rarely, in Oregon.

Rubber rabbitbrush
Ericameria nauseosa (Pallas ex Pursh)
G. L. Nesom & G. I. Baird

There are twenty-two varieties of *Ericameria nauseosa (*formerly *Chrysothamnus nauseosus)*. Add several other species also known as rabbitbrushes, and one ends up with a taxonomic jumble. We'll consider the plant shown here as representative and leave it at that. The rubber rabbitbrush has clusters of rayless, bright yellow flowerheads and frosted blue-green, linear leaves. The sticky latex-like sap is white, explaining the name "rubber rabbitbrush." Meriwether Lewis was confused by the rabbitbrushes, the likes of which he had never before encountered—not surprising for they were then unknown to science—and he collected two specimens of this plant. The plant, a late bloomer, is found throughout the Great Plains and western states, north to adjacent Canadian provinces and south to Mexico.

Green rabbitbrush
Chrysothamnus viscidiflorus (Hook.) Nutt.

The green rabbitbrush is less striking than the rubber rabbitbrush (on the preceding page) although it is also part of the mid- to late summer landscape in much of the West. There are five different varieties of *Chrysothamnus viscidiflorus*; three grow in Idaho. The differences between varieties are relatively minor. All are crowded, clustered plants with woody stems and variably hairy, narrow green leaves. The stems, leaves and flowers often feel sticky, explaining the name *viscidiflorus*. Narrow, discoid, brush-like flowers characterize rabbitbrushes in general.

Black-hairy prairie-dandelion
Nothocalaïs nigrescens
(L. H. Hend.) A. Heller

As their common name "dandelion" suggests, the mountain dandelions (*Agoseris* spp) and several other yellow-rayed plants have inflorescences that resemble those of the common dandelion. They have other common characterics: milky juice, taproots and dandelion-like pappuses, and they lack disk flowers.* Despite this plant's common name, prairie-dandelion, it is a mountain plant that grows only near the common borders of Wyoming, Montana and Idaho. The Latin species name, *nigrescens,* means "turning black" for the dark markings on the involucral bracts (giving it another common name, speckled false dandelion), the pointed leaves that cup the flower parts.

---

* Some of these dandelion-like species have been reclassified several times. The plant shown here was originally published as *Microseris nigrescens*, joining a few other plants in the genus *Microseris*. Some of these have recently been reclassified as *Nothocalaïs,* a name coined by Asa Gray (1810–1888) of Harvard College, a man who made his reputation naming plants that others collected. Needless to say, reclassifications can be confusing, not only to lay plant-lovers, but to many botanists as well.

Calaïs was a minor Greek god, son of Borealis, the north wind. The prefix *notho-* means "false," suggesting that the genus was similar to, but not the same as *Calaïs* (today's genus *Uropappus*).

Low hawksbeard
***Crepis modocensis* Greene**

The low hawksbeard has only ray flowers. Its leaves are deeply incised, long-stemmed and pinnate (feather-like). The Greek word *krepis* means "sandal," used by Theophrastus for a similar plant. The species name, *modocensis*, refers to Modoc County in California. Other *Crepis* species also grow in Idaho; the form of their deeply serrated leaves helps to identify them. A tendency to crossbreed may make identification difficult. *Crepis modocensis* is native to the three western coastal states and east through the Rocky Mountains.

The etymology of the name "hawksbeard" is obscure. Possibly because the seed pappus is bristly it suggested the "mustache" feathers of nighthawks (nightjar family, Caprimulgidae).

Western hawkweed
***Hieracium scouleri* Hook.
var. *albertinum***
(Farr) G. W. Douglas & G. A. Allen

Hawkweed's generic name was derived from the Greek *hierax* for "hawk." The half-inch-wide flowerheads are without disk florets. All parts of the plant except the flowerhead itself are hairy, and the sap is milky. The western hawkweed blooms from midsummer on, as high as the subalpine zone. Interestingly, Gregor Johan Mendel (1822–1884) attempted to repeat his genetic experiments—originally carried out with pea plants—using a species of *Hieracium*. He did not know that hawkweeds may reproduce asexually by a process known as apomyxis. His results were so inconsistent that he gave up on further plant experimentation. John Scouler (1804–1871) was a naturalist who visited the Northwest briefly in 1825-26.

Common eriophyllum
***Eriophyllum lanatum* (Pursh)
J. Forbes**

Various common names including "woolly sunflower" and "Oregon sunshine" have been suggested for this attractive composite, but it is usually known simply as an eriophyllum, (pronounced eri-OFF-illum). It is distributed widely in the West, and a dozen or so varieties are recognized. Ours is var. *integrifolium* (Hook.) Smiley, characterized by entire (unlobed) leaves and seven or eight wide rays. The leaves are covered with fine hairs giving them a silvery color, explaining the species name, *lanatum* ("woolly"). Eriophyllums are a meadow plant that grow to subalpine elevations, blooming from late spring on. Lewis and Clark saw eriophyllums above their camp on the Clearwater River near present-day Kamiah, Idaho, where they gathered two specimens on June 6, 1806.

Rocky Mountain Canada goldenrod
Solidago lepida DC. var. *salebrosa* (Piper) Semple

The goldenrod shown here is found from British Columbia, east to Saskatchewan and south to Arizona and New Mexico. It grows to fairly high elevations from midsummer on, usually in open meadows. Its varietal name, *salebrosa*, means "rough" for the surface of the plant's leaves. If you examine the plant's bloom closely, you'll find that it is made up of hundreds of tiny composite flowerheads, and that each of these little heads has ray flowers less than a tenth of an inch in size. The generic name, *Solidago*, was derived from two Latin words, *solidus* meaning "complete" and *ago* for "I make whole," because the European goldenrod (*Solidago virgaurea*) was, in times gone by, valued as a vulnerary—a substance able to heal external wounds.

Mountain goldenrod
Solidago multiradiata Aiton

The rays are more obvious on the mountain goldenrod than they are on the goldenrod shown above. This is a summer-blooming plant, often seen along our trails where the ground is moist. As its common name suggests, it is only encountered at higher elevations; in Idaho it grows at least as high as treeline. In common with many of our alpine plants, this one occurs at progressively lower elevations as one goes farther north in its range—Canada, Alaska and Siberia. There are several varieties of mountain goldenrod, although this is the only one found in Idaho. While the common name "mountain goldenrod" describes the plant as it grows in the West, it is not wholly accurate because the same plant also grows in Labrador and southeastern Canada at sea level.

Tundra hymenoxys
Tetraneuris grandiflora
(Torr. & Gray) K. F. Parker

The tundra hymenoxys is a true alpine plant. Because of its hairy leaves and stems it is sometimes called "the old man of the mountains." Given its alpine surroundings, its large flowerhead and plump central disk, it is easy to recognize. The plant grows in the high mountains of Idaho, Montana, Utah, Wyoming and Colorado. There has been some confusion about the name. Known variously as *Hymenoxys* and *Tetraneuris*, the latter generic name is now preferred for this plant and for the related stemless hymenoxys shown below.

Stemless hymenoxys
Tetraneuris acaulis **(Pursh) Greene**

The stemless hymenoxys is less well known than its alpine relative, although the flowerheads, one to a stem, with large central disks, suggest a kinship. This is a variable plant—five varieties have been described; two grow in Idaho (the plant illustrated seems to be var. *acaulis*). Its rays are wide and their number varies; occasionally the plants are rayless. They prefer rocky ground and grow, as one or another variety, throughout the West from Alberta, east to Kansas and the Dakotas, west to Nevada and California and as far south as Texas. The Latin term *acaulis* means "stemless," referring to the leaves, not to the inflorescense.

Alpine hulsea
Hulsea algida A. Gray

The alpine hulsea is an attractive yellow composite that has adapted to living at treeline and higher. It grows in well-protected crevices on talus slopes where it flowers in mid- to late summer. Serrated leaves, a thick stem and a yellow flowerhead suggest a common dandelion, but a closer look shows that its leaves are thick and the flowerhead has both ray and disk flowers. Its sticky leaves, like those of many composites, give off a pronounced aromatic odor when crushed. This is the only hulsea found in Idaho. (A dwarf species, *Hulsea nana,* is native to the Cascade and Sierra Nevada ranges.)

Hulseas were named for United States Army physician and botanist Dr. Gilbert White Hulse (1807–1883). The Latin word *algida* means "cold," a reflection of the plant's alpine environment. "Pacific alpinegold" has been suggested as a standardized common name for this plant.

Rocky Mountain groundsel
Packera streptanthifolia (Greene) W. A. Weber & A. Löve

The Rocky Mountain groundsel is commonly found growing in mid- to late summer on the spongy, moist soil of high-elevation bogs and fens. The plants have a cluster of basal, ovoid, serrated leaves and long, whippy, naked stems. They are tall so it is difficult to show the entire plant photographically. As with many former senecios, it is now in genus *Packera.** (This one previously was classified as *Senecio cymbalaroides.*) The species name, *streptanthifolia,* was derived from the Latin and means "crown-like leaf" from the leaves' serrated edges. Rocky Mountain groundsels are found in most of the western states and Canadian provinces and as far north as Alaska.

---

* The name *Packera* honors the Canadian botanist, Dr. John G. Packer (1929–), of the Department of Botany at the University of Alberta. The genus *Senecio* has been divided into several genera primarily on the basis of chromosomal differences. These taxonomic changes are a source of confusion for those used to the former generic names.

Split-leaf groundsel
Packera dimorphophylla (Greene) W. A. Weber & A. Löve var. *paysonii* (T. M. Barkl). D. K. Trock & T. M. Barkl.

The split-leaf groundsel, shown here growing in the shaded surroundings of a montane forest, has a rosette of somewhat thickened, almost succulent leaves and a few clasping stem leaves, surmounted by a loose cluster of flowerheads bearing from seven to nine rays and a rounded central disk. The species name, *dimorphophylla,* means "leaves having two shapes" (i.e., some round, some elongated) as suggested in the photograph. The plant is found in Idaho, Montana, Wyoming, Utah and Nevada.

Dwarf arctic groundsel
Packera subnuda (DC.) Trock & T. M. Barkl

This little groundsel (previously classified as *Senecio cymbalaria*) is a late-blooming alpine plant, whose prostrate configuration is typical of many alpine plants. Tiny toothed leaves and few-rayed flowerheads help with its identification. It occurs in British Columbia, Alberta and south to California and Nevada, as well as in Idaho. The species name, *subnuda*, refers to the relatively long, nearly leafless stem.

Rocky alpine groundsel
Packera werneriifolia (A. Gray) W. A. Weber & A. Löve

While this little packera is occasionally seen at lower elevations, it is primarily a high-altitude plant. It was classified as a *Senecio* until recently. The flowerhead is rayed, although the rays are extremely small. Its ovoid gray-green leaves and purplish involucres help to identify the plant. It is found in all our Rocky Mountain states and west to Arizona and California.

Streambank groundsel
Packera pseudaurea (Rydb.) W. A. Weber & A. Löve

Several similar plants, some formerly in the genus *Senecio*, grow in moist substrates that range from wet meadows to standing water. They are tall, long-stemmed plants. Most have serrated leaves and small clustered flowerheads, prominent central disks and small or absent ray flowers. This plant is an example. The streambank groundsel is a common plant found in most of the central and western states and Canadian provinces.

Alkali-marsh ragwort
Senecio hydrophilus Nutt.

The photo at the left is another plant that grows on moist ground and in the still water of bogs, fens and moist meadows. The alkali-marsh ragwort shown here seems to be the most common of the moisture-loving senecios. It grows in the western coastal states, east to South Dakota and in the intervening states. The plants are characterized by smooth (occasionally serrated), relatively large basal leaves that become smaller and more pointed as they ascent the lower stem. The hollow stem is stout and maybe a dark reddish-purple. The flowerheads sometimes have small petals, but more commonly are discoid as in the plants shown here.

Jane Lundin

Dwarf mountain ragwort
Senecio fremontii Torr. & Gray var. *fremontii*

The dwarf mountain ragwort was first collected by John C. Frémont (1813–1890), army engineer, explorer, adventurer and eventual politician, on one of his several western expeditions during the 1830s. The plant was described subsequently by academic botanists John Torrey (1796–1873) and Asa Gray (1810–1888) in their *Flora of North America* (1843). Our photograph shows the plant's typical appearance as it is usually found growing among boulders on a subalpine talus slope. It is characterized by thick, almost succulent, toothed leaves and small clusters of few-rayed flowerheads. Several varieties have been described, but this one is the only one that grows in Idaho. The species, as one variety or another, ranges from the high mountains of the two western provinces of Canada and the adjacent American states as far south as California and New Mexico.

Tall ragweed
Senecio serra Hook.

The tall ragweed (or ragwort or butterweed) is a common western meadow plant that blooms from early to midsummer as high as the subalpine zone. Great numbers of small ray flowers are gathered into crowded clusters. The species name, *serra* ("saw"), refers to its narrow, stemless and (usually, but not always) toothed leaves. The tall ragweed grows in Idaho and all the adjacent states, and spills over into Colorado and California. The related arrow-leaf ragwort, *Senecio triangularis* Hook., is an almost identical plant except for its wider, triangular-shaped, serrated leaves. It has much the same distribution in the lower forty-eight states, although it grows farther north in the adjacent Canadian provinces and to Alaska.

Ballhead ragwort
Senecio sphaerocephalus Greene

This plant, common at higher elevations, is one of the earliest blooming of the rayed composites. It is easily identified by its rounded, clustered yellow flowerheads that give the plant both its common and scientific names (its species name, *sphaerocephalus*, from the Latin, means "round head"). Black-tipped bracts (the leaflets that cup the flowerhead) are another distinguishing feature that help to identify the plant. It occurs in Idaho, Montana, Wyoming, Utah and Nevada.

Slender tarweed
Madia gracilis (Sm.) D. D. Keck

Tarweeds are named for their aromatic tar-like odor; this, and the slender tarweed's unusual twice-notched rays, help to identify it. The generic name, *Madia,* was derived from a related plant's (*Madia sativa* Molina) name. The latter plant also grows in Chile, where it is known as *madi.* The slender tarweed is native to all the northwestern states, as well as Utah, Nevada, California and British Columbia. Oddly and inexplicably, an isolated population is found in Maine.

Mountain tarweed
Madia glomerata Hook.

This odd little subalpine plant's flowerheads have small clusters of irregular ray flowers and, as shown here, variably large hairy disks. The hairy flower parts exude sticky drops of fluid as seen in the photograph. These give off a strong tar-like odor. It's not unusual for mountain tarweeds to grow in great numbers along our trails at montane and subalpine elevations, and the air becomes redolent with the plants' distinctive odor. This tarweed grows throughout the West, in most of the provinces across Canada and north to Alaska.

Stemless goldenweed
Stenotus acaulis (Nutt.) Nutt.

Although the stemless goldenweed was included in the genus *Haplopappus* (and *Aplopappus* prior to that) for many decades, it has recently been returned to Thomas Nuttall's original classification. He apparently derived the generic name, *Stenotus,* from the Greek *stenos,* a word that means "narrow," presumably for the shape of its leaves. Its species name, *acaulis,* means "without a stem," referring to the leaves. The plant grows in Idaho, Montana, Oregon, Colorado and south to California, from the foothills to the alpine tundra, where this plant was photographed.

Woolly goldenweed
Stenotus lanuginosus (A. Gray) Greene
var. *andersonii* (Rydb.) C. A. Morse

The woolly goldenweed (formerly *Haplopappus lanuginosus* var. *andersonii*) is commonly seen growing in our mountains from montane to subalpine zones, typically preferring rocky or gravelly soil. It is characterized by basally clustered, soft, narrow and rather hairy leaves. The showy flowerhead has a prominent disk and wide, deep yellow rays. It occurs in Idaho, Washington, Oregon and Nevada. A variant, var. *lanuginosus*, occurs in southwestern Idaho, central Washington, northeastern California and northwestern Nevada.

Yellow mule's-ears (left)
Wyethia amplexicaulis (Nutt.) Nutt.

The yellow mule's-ears bears some resemblance to the arrowleaf balsamroot (page 34). Both are large plants with showy blooms, although this one prefers wet meadows rather than dry hillsides. Unlike the balsamroot, it has stemless leaves (*amplexicaulis* means "stem-clasping"). Further, its leaves are shiny, sometimes described as having a varnished appearance. Yellow mule's-ears bloom in the spring, usually a week or so after the white mule's-ears, shown below. Boston businessman Nathaniel Wyeth (1802–1856) collected both wyethias (and other plants as well) for botanist Thomas Nuttall, while in today's western Montana in 1833.

The two species of *Wyethia* shown here grow only in Idaho and neighboring states. Interestingly, when white and yellow wyethias grow side by side, they may hybridize as a pale yellow-flowered form, *Wyethia* x *cusickii* Piper (upper right).

White mule's-ears (left)
Wyethia helianthoides Nutt.

This species name, *helianthoides*, means "sunflower-like"—which seems strange, as one would think that the yellow mule's-ears would have been given that name. Showy, large, spring-blooming, white wyethias are found along seasonal streams and in moist meadows, often in vast numbers. Massed wyethias in bloom are a striking sight and an irresistible subject for photographers.

Prickly lettuce (left)
Lactuca serriola L.

The prickly lettuce is a common weed imported from Eurasia. It is now found almost everywhere growing on disturbed ground where it blooms from mid- to late summer. The plant's insipid yellow flowerhead lacks disk flowers. Prickly, lobular leaves grasp a woody stem with small "ears." Prominently pointed buds and milky sap confirm its identification. *Lactuca serriola* is believed to be the ancestor of our edible lettuce, *Lactuca sativa*.

Blue lettuce (left)
Mulgedium oblongifolium (Nutt.) Reveal

This blue lettuce (formerly *Lactuca pulchella*) is the showiest of our several native wild lettuces. Despite its attractive flowerhead, it is considered a weed. Its leaves, growth habit and distribution are similar to those of the prickly lettuce. This plant is only occasionally encountered as high as the montane zone, although it is a common weed lower down.

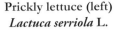
Jane Lundin

Yellow salsify (above)
Tragopogon dubius Scop.

As its species name, *dubius*, suggests, the yellow salsify is part of a confusingly interbred genus, one that includes the similar, blue-flowered salsify, *Tragopogon porrifolius,* also known as the oyster plant. Both were introduced from the Old World for their edible roots (although this plant's stringy roots would hardly seem edible). They now grow almost everywhere in the United States except the Deep South. *Tragopogon* means "goat's beard," from the plant's large, dandelion-like, feathery seed head.

Common tansy
Tanacetum vulgare L.

Tansy is a Eurasian perennial that now grows everywhere in the Northern Hemisphere. It is sometimes seen in Idaho at higher elevations, always near settled places. The tansy's button-like flowers and camphor-like odor are unique, so there will be no problem with identification. Not only is the plant ornamental, but it is also an effective vermifuge. Tansy teas and extracts were used medicinally for other conditions, although with some risk and questionable benefit. Tansies also contain the insect repellent pyrethrum, and their leaves were formerly used to wrap meat to prevent spoiling, and were added to winding sheets to deter worms, supposedly explaining the derivation of *Tanacetum* (and "tansy") from the Greek *athanasia*, a word that means "no death" (i.e., "immortal").

Pineapple weed
Matricaria discoidea DC.

Meriwether Lewis found specimens of this plant, native to Idaho, growing along the Clearwater River on June 12, 1806. He described it as "a small plant of an agreeable sweet scent; flowers yellow." The crushed plant gives off an odor close to that of a pineapple, explaining its common name. Related to the European plant used to make camomile tea, ours has been used for the same purpose. It is a common, non-aggressive garden and border weed found in almost every state and province. The pineapple weed grows in our mountains, usually on disturbed ground, to mid-elevations where it blooms from midsummer on. The name *Matricaria* is said to imply "mother care" for the plant's supposed value in treating uterine conditions.

Large-flowered brickellia
Brickellia grandiflora
(Hook.) Nutt.

The brickellias are plain cousins of the showier joe pye-weeds (*Eupatorium* spp.); they are not plants that one would collect for an ornamental garden. You will see them growing at high elevations in our mountains and, as one species or another, at lower elevations elsewhere west of the Mississippi River. The plants' pale yellow to white flowerheads lack rays and appear to be squeezed together and held in place by the involucral bracts. Prominent delta-shaped leaves also help to identify this species. The plants bloom late, from midsummer on. The genus name, *Brickellia*, honors botanist and physician John Brickell (1748–1809) of Savannah, Georgia. Several similar species of *Brickellia* are also native to Idaho.

Jane Lundin

Dusty maiden	Alpine chaenactis	Evermann's chaenactis

Dusty maiden
Chaenactis douglasii
(Hook.) Hook. & Arn.
var. *douglasii*

The generic name *Chaenactis* was derived from the Greek *chaino* meaning "to gape" and *aktin* meaning "ray," for the wide-mouthed flowers on the periphery of the disk. One might not recognize this frilly-leaved plant as an Asteraceae, but the protruding forked styles are a tipoff. The florets are white to pale pink, their color accentuated by the pink styles. The dusty maiden grows to fairly high elevations and blooms—often in great numbers—on dry slopes. Dull gray-green (glaucous), frilly leaves explain the common name, "dusty maiden." (The name "pincushion" has also been proposed for plants in this genus, apparently to avoid duplication with other dusty maidens.)

Alpine chaenactis
Chaenactis douglasii
(Hook.) Hook. & Arn.
var. *alpina* A. Gray

The alpine chaenactis is a small perennial that blooms toward summer's end, among the rocks of talus slopes near treeline and above. It resembles a smaller version of *Chaenactis douglasii* var. *douglasii,* differing in its white flowerheads and bare stems. The differences between the two plants were formerly deemed sufficient to allow each its own species designation; now, however, they are usually classified as varieties of the same species. Darwin would have been pleased to see how the two plants have evolved in different ecological settings from a common ancestral form. While this variety is restricted to Idaho and surrounding states (less Washington and Nevada), var. *douglasii* is widespread in the far West.

Evermann's chaenactis
Chaenactis evermannii Greene

Evermann's chaenactis (or pincushion) is an uncommon plant. It grows on rocky ground at high elevations, often in the shelter of the boulders of talus slopes. Its leaves are flattened compared to those of the other *Chaenactis* shown on this page. The woolly flowerheads are another distinguishing feature. Evermann's chaenactis grows only in the mountains of central Idaho.

Rosy pussytoes (left)
Antennaria rosea Greene

Pussytoes are not particularly attractive, but they are so numerous that one cannot help but notice them—this one especially, for its reddish hue. A rosette of basal leaves comes off a stem surmounted by a cluster of small flowerheads about the size of a house cat's digital pad, whence their common name. After the "toes" open, one can see that each flowerhead is made up of many tiny flowers. There are several varieties of *Antennaria rosea*—we will not attempt to identify this one further. The name *Antennaria* was apparently derived from the resemblance of the flower's pappus to an insect antenna.

Rocky Mountain pussytoes
Antennaria media Greene
(below left)

The Rocky Mountain pussytoes, shown here in flower, grows on alpine and arctic tundra in our western mountain states and provinces, north to Alaska and south to California, Arizona and New Mexico. Its tiny oval leaves are covered with fine hairs giving them a color more gray than green. It is commonly found on rocky ground where it is nurtured by the retained heat of the sun.

Antennaria sp. (right)

The photo shows the typical appearance of grouped pussytoes (most likely *Antennaria umbrinella* Rydb.). There are many antennarias; the species shown on this page are not hard to identify, but others require an appropriate reference source for help with their classification.

Pearly-everlasting (above)
Anaphalis margaritacea
(L.) Benth. & Hook. f.

Anaphalis is the ancient Greek name for a similar plant. The name *margaritacea* echoes its common name, "pearly." Each of its round, white flowerheads has a characteristic diffuse black dot on the surrounding involucre. The only member of the genus, it is related to *Antennaria* and grows in most of North America. It is often used in dried flower arrangements. Supposedly the pearly-everlasting was the first North American herb to be cultivated in Europe, because of purported medicinal value.

Hooker's pussytoes
Antennaria racemosa Hook.

While this plant's forked styles serve to identify it as a member of the aster family, its smooth green leaves and loose clusters do not immediately suggest that it is an antennaria. Smooth-leaved plants arise from creeping stems (stolons). The flowerheads are rather larger and looser than those of other antennaria. Hooker's pussytoes are most often found in the open shade of wooded areas. Look for them in central Idaho and in adjacent states of the Northwest, south to Nevada and California and in the two adjacent westernmost provinces of Canada.

Western snakeroot
Ageratina occidentalis
(Hook.) R. M. King & H. E. Rob.

This plant, also known as western boneset, was until recently classified as a *Eupatorium,* a genus that includes the common joe pye-weeds. This, and species shown on neighboring pages, are rayless composites. Protruding forked styles, a distinguishing feature of composites in general, give the flowers a feathery appearance. Western boneset is relatively uncommon, favoring subalpine to alpine cliffs and other rocky surroundings. The name *Ageratina* was that of an unknown ancient Greek plant and means "everlasting," from the suffix *a* for "not" and *gera* for "old" (for its long-lasting flowers). David Douglas collected this species on the Lewis and Clark River (today's Snake River). It grows also in the coastal states and in Nevada and Utah.

Lesser burdock
Arctium minus Bernh.

Burdocks are usually seen after the flowers are dried, forming the burrs beloved by children for their tendency to stick to whatever they are thrown at. In this photograph one can see an early flowerhead characterized by tiny purple rays and prominent, white-tipped, forked styles. The tips of the bracts end in the inwardly curving hooklets that fasten onto whatever passes by. (These served as the inspiration for Velcro.) This, and the larger *Arctium lappa*—the greater, or edible, burdock—are both natives of Europe. Both weeds have spread throughout the United States and lower provinces of Canada, probably brought to the New World by the earliest settlers. The lesser burdock grows near settled places in our mountains at least as high as the montane zone.

Hooker's thistle
Cirsium hookerianum Nutt. (left)

Hooker's, or white thistle, was given the name *hookerianum* by Thomas Nuttall in 1841 to honor William Jackson Hooker (1785–1865), professor of botany at Glasgow and later director of England's Royal Botanic Garden at Kew. The plant grows at high elevations in the coastal ranges of British Columbia, the Cascade Range in Washington and the Rocky Mountains states of Idaho, Montana and Wyoming. Its white flowerhead makes identication easy. *Cirsium* is a Greek word for a "knot of veins," a condition that thistles were used to treat in the distant past.

Elk thistle
Cirsium foliosum (Hook.) DC. (below left)

The elk thistle blooms in early summer along mountain streams and in wet meadows. The young plants are eaten by elk and bears; the peeled stems are edible for humans. The plants are easily identified by their size (it is our largest native thistle), by its many prickly pinnate leaves (*foliosum* means "leafy"), and by bracts (specialized leaves that cup the flower parts) that extend well above a white or pinkish flowerhead that turns brown as the plant matures. Both this plant and Hooker's thistle were gathered by Thomas Drummond (1780–1835), a Scot who collected plants in western America. (Some believe that this plant should be classified as *Cirsium scariosum*, the meadow thistles. We will retain the above classification.)

Jackson Hole thistle
Cirsium inamoenum (Greene) D. J. Keil (below)

The Jackson Hole thistle (*Cirsium subniveum* is an earlier scientific name) grows along roads and trails, blooming in midsummer. The plants have a bush-like appearance, growing year after year in the same location. They have hard, serrated, spiny leaves and pale pinkish or lavender thistle-like flowers borne on branching stems. It grows in Idaho and contiguous states as well as in California.

Spotted knapweed
Centaurea stoebe L.

The spotted knapweed grows through-out the United States. Difficult to eradi-cate, it is among our most troublesome of the nonnative weeds. Identify it by the spotted involucre that cups the flower parts. It is found in our mountains as high as the montane zone.

Yellow star thistle
Centaurea solstitialis L.

The yellow star thistle is another thor-oughly noxious Eurasian *Centaurea* now found throughout most of the United States and Canada. Deep rooted and rapidly spreading, it is especially com-mon in the Snake River foothills of western Idaho.

Cornflower
Centaurea cyanus L.

The cornflower (also known as bach-elor's button) is a favorite wildflower in Europe, and is the national flower of Poland. It has established itself as a foothill plant in Idaho and elsewhere. For the present, at least, it seems to be a relatively benign weed in Idaho.

Canadian thistle (left)
Cirsium arvense (L.) Scop.

The Canadian thistle, like the spotted knapweed, is a noxious plant that now grows in all but a few southern states and in all the Cana-dian provinces. A tall plant with pink to light purple flowers, it is not easily missed. Like the knapweed it is a deeply rooted, perennial plant that is difficult to eradicate. Canadian thistles commonly grow on disturbed ground, usually along roads and railroads. Recently we have seen it growing high in the montane zone, well away from a populated area.

Common yarrow
Achillea millefolium L. var. *millefolium* (left)
Achillea millefolium L. var. *alpicola* (Rydb.) Garrett (below left)

The common yarrow is a circumboreal plant that resembles members of the carrot family (Apiaceae)—look closely, however, and you'll see that each "flower" has, in addition to tiny little ray flowers, a disk made up of minute florets. The yarrow is a survivor; it grows all through the Northern Hemisphere, blooming from late spring into the fall. If we look at a map showing its distribution, we see that the yarrow, as one variety or another, grows in every state, in every Canadian province, in Greenland and in all the arctic islands between.

The name *Achillea* comes from the belief that Achilles used yarrows to treat his companions' wounds. This is understandable for, as with many members of the Asteraceae family, it has an aromatic, medicinal odor. The plant and its relatives have been used medicinally for millennia by many cultures, although it has no scientifically proven therapeutic value. The species name, *millefolium*, describes the plant's finely divided leaves. As with many widely distributed species, there are many—a dozen or so—varieties.

The lower image shows a subalpine/alpine variety that grows at high elevations in Idaho and in the mountains of other western states and provinces. Its varietal name, *alpicola*, from the Latin, means "of high mountains." The plant's foliage is densely covered with hairs that serve the same purpose as in animals, (i.e., as protection from the cold of the high elevations at which the plant lives).

Lewis and Clark collected the common yarrow on May 20, 1806, in today's northern Idaho.

Big sage
Artemisia tridentata Nutt
var. *vaseyana* (Ryd.) Beetle (left)

Several species of *Artemisia* grow in our mountains. The big sage is the plant we think of as the western "sagebrush." It is ubiquitous, growing in most of our western states. Like all artemisias, it is highly aromatic with small three-lobed leaves that give the big sage its species name (*tridentata* means "three-toothed"). There are at least six varieties of big sage. The most common variety of sagebrush at higher elevations, var. *vaseyana,* is recognizable by its prominent, high-standing, flower-bearing spike.

Silver sage
Artemisia cana Pursh (above right)

The silver sage is an attractive plant with blue-gray leaves that explain its common name. While it grows as high as the montane zone, we have not encountered it at higher elevations. Its leaves are mostly entire (i.e., unlobed), but occasionally they have three-lobes, explaining why it was formerly classified as a variety of big sage even though its appearance is quite different. Further, unlike the other, it prefers moist ground. *Artemisia cana* is the least common of the three plants shown here.

Louisiana sage (right and left)
Artemisia ludoviciana Nutt. var.
ludoviciana

The Louisiana sage (also prairie sage, white sage and western mugwort) is the second most common of Idaho's artemisias. It grows throughout the United States and Canada. Several varieties are recognized. The variety *Artemisia ludoviceana* var. *candicans* (Rydb.) Keck (right) is not seen often because it is an alpine plant. Its leaves are more divided, smoother and greener than those of the common variety found at lower altitudes (left). Louisiana sage, like other artemisias, has a strong herbal odor that is helpful in identifying members of this genus.

Barberry Family (Berberidaceae)

The Barberry family consists of thirteen genera and 660 species. While most are in the northern temperate zone, others are scattered throughout the world in a haphazard fashion, suggesting—along with certain plant characteristics—that they are one of the less specialized of the flowering plants. The best known North American species are the eastern mayapple (*Podophyllum peltatum*) from which the anti-tumor medication podophyllin is obtained and, in the West, the several species of Oregon grape (species of *Berberis*). The common barberry (*Berberis vulgaris*), a red-berried hedge-plant, is a native of Europe. Several other barberries, e.g., the Japanese barberry (*Berberis thunbergii*) as well as the various Oregon grapes, are used as ornamentals in hedges and as ground covers. Most plants in this family are woody, sometimes brambly, shrubs. Some are evergreen. The flowers have four to six petals, although because they are joined, the petals can be hard to count. The origin of the word "berberis," from which "barberry" was derived, is unknown; the resemblance to "berry" apparently is fortuitous.

Creeping Oregon grape
***Berberis repens* Lindl.**

The creeping Oregon grape (or mountain holly) grows in open woods where it spreads by woody rhizomes (*repens* means "creeping"). Deep green to rusty-red holly-like leaves and distinctive flowers and fruit make it easy to identify. Blue berries appear in mid- to late summer and, while edible, are best used for preserves. The plant grows in all the western states and provinces, as well as in scattered locations further east. It is sometimes also classified as *Mahonia repens*, honoring Irish-born Philadelphia horticulturist Bernard M'Mahon (1775–1816). M'Mahon had access to Lewis and Clark's plant specimens and grew this plant from their seeds, probably ones collected in northern Idaho in 1805 or 1806.*

---

*The explorers brought back specimens of two other Oregon grapes—shiny-leaved *Berberis aquifolium* and dull-leaved *Berberis nervosa,* collected at today's The Dalles in Oregon; both plants, unlike the creeping Oregon grape, are bushes, often used in landscape gardening. Their flowers and fruit are similar to those shown here. These plants also grow in Idaho, but *Berberis repens* is the species found in our mountains.

Boraginaceae

Borage, or Forget-me-not Family (Boraginaceae)

The Borage family's name was derived from an attractive, blue-flowered European plant, *Borago officinalis*, found in the United States only as an imported ornamental. Its name possibly comes from the Latin *burra* meaning a "rough garment" or "coat," referring to the plant's hairy leaves and stem. As borage is unfamiliar to most Americans, Forget-me-not is often used as the family common name. There are 117 genera and 2,435 species in the family; 90 or so species grow in the American Northwest. Many look so much alike that identification of individual plants can be difficult, depending more on seed (nutlet) properties than on plant appearance. Family characteristics include alternate bristly leaves and loose clumps of flowers borne on a stem that in some species seems to unroll (i.e., they're scorpioid) as the flowers open. The flowers are often blue, but may also be white, pink or yellow. They are radially symmetrical. Five petals (occasionally four) are joined at the base to form a tube. The ovary is four-parted; each part forms one nutlet or seed. In some species the nutlets have barbed spines that stick to the fur of passing animals and to hikers' socks and shoelaces, explaining why "stickseed" is a common name, especially those in genus *Hackelia*. Most of the Boraginaceae are herbaceous (i.e., non-woody). Some are used as ornamental plants and this is the family's chief economic importance. Ornamental Boraginaceae include *Myosotis* (forget-me-nots), *Heliotropium* (heliotropes), *Mertensia* (bluebells) and others.

Meadow forget-me-not
Hackelia micrantha (Eastw.) J. L. Gentry

The meadow forget-me-not, shown here considerably magnified, is also known as the small-flowered (or false) forget-me-not, blue stickseed, or simply as a hackelia. Its nutlets are armed with small prickles that cling tenaciously to the fur of animals or the clothing of passersby, explaining the common name "stickseed." The plants bloom in late spring or early summer, growing at least as high as treeline and forming blue patches on mountain slopes. The tall, stout-stemmed plants have furry, lanceolate leaves. Their small flowers are borne in loose clumps. Each petal has a distinguishing longitudinal fold at its base. While the flowers are usually blue, occasionally they are white or pink. The species ranges from British Columbia and Alberta, south to California, Nevada, Utah and Colorado. The name *Hackelia* honors Czech botanist Joseph Hackel (1783–1869).

61

Jane Lundin

Spreading stickseed
Hackelia patens (Nutt.) I. M. Johnst.

The spreading stickseed has the five-pet-aled flowers typical of those seen in other hackelias as well as in several other genera in the family Boraginaceae. Because their flowers resemble the true forget-me-nots (*Myosotis* spp.), they are sometimes called "wild (or false) forget-me-nots." The name "spreading stickseed" has been suggested as a standardized common name for this plant. It is not uncommon although it is less widely distributed than the small-flowered stickseed shown on the preceding page. The petals often have blue markings. These range from a pale blue wash to discrete blue marks. The spreading stickseed grows in Idaho (where it is common), Montana, Wyoming, Utah and Nevada, and uncommonly in Oregon.

Slender popcorn-flower
Plagiobothrys tenellus
(Nutt. ex Hook.) A. Gray

Several genera in the borage family are made up of very small plants. This plant's flowers are only about an eighth of an inch in diameter, and the plant may stand no more than three inches high. It ranges from British Columbia, south to Baja California and east to Idaho, Utah and Arizona. The name *Plagiobothrys* was derived from two Greek words and means "obliquely pitted," referring to the appearance of the plant's nutlet. The term *tenellus* is Latin for "slender." Meriwether Lewis collected this new-to-science plant at The Dalles, in present-day Oregon, on April 17, 1806, although his specimen was overlooked; it was found and classified later by Thomas Nuttall.

Waterton Lakes cryptantha
Cryptantha sobolifera Payson

The genus *Cryptantha* includes about forty species, all found west of the Mississippi River; approximately fifteen species grow in Idaho, mostly at lower elevations. The plant shown here, also known as the alpine cryptantha (or alpine cat's eye) is an exception, for it is a true alpine plant. It is small, standing one to five inches high. Its leaves, stems and calyces are notably bristly, and its tiny, clustered, five-petaled white flowers are borne on relatively long stems that arise from a basal gathering of lanceolate leaves. The species name, *sobolifera,* from the Latin, means "sobole-bearing." (A sobole is a shoot or sprout— i.e., a sucker—that grows at the base of a plant.)

Arctic alpine forget-me-not
Eritrichium nanum (Villars) Schrader

This forget-me-not is a striking little alpine plant whose bright blue blossoms stand out vividly against drab mountain tundra. The plants form matted "cushions" made up of tiny, tightly clustered leaves. This lovely little plant was photographed above treeline in our White Cloud Range. It is found at high elevations throughout the Rocky Mountains as far south as New Mexico, north to Alaska and in the mountain ranges of Europe and Asia.

Asian forget-me-not
Myosotis asiatica (Vesterg.) Schischk. & Serg.

The Asian forget-me-not is a circumboreal plant that prefers moist montane, subalpine and alpine meadows. It grows in our Northwest, south to Colorado and north to Alaska. Its small, five-petaled, bright blue flowers, similar to those of many other plants in the borage family, and wide lanceolate leaves identify the plant. *Myosotis* was derived from two Greek words meaning "mouse ear," used in the past for a now unknown plant. Plants in this genus are considered to be the true forget-me-nots.

Alpine bluebell
Mertensia alpina (Torr.) G. Don

The alpine bluebell, in common with many alpine plants, is small, standing only four or five inches high. It is found only in Idaho, Montana, Wyoming, Colorado, and in New Mexico where it is rare. It is also uncommon in Idaho, reportedly found only in the higher mountains close to the Idaho-Montana border in Fremont County.* It is an attractive little plant that may be recognized by its small size and flaring petals and by its presence on alpine tundra.

---

*The plant shown here was photographed in adjacent Beaverhead County, Montana.

Boraginaceae

Ciliate bluebell
Mertensia ciliata
(James ex Torr.) G. Don

The ciliate (or mountain) bluebell's species name, *ciliata*, means "fringed" for the fine hairs that can be seen along the edges of back-lighted leaves. The plants grow at least as high as treeline in moist meadows and along streambanks, often forming "rivers" of green and blue. The flowers are deep blue when shaded, and a lighter color when growing in full sunlight. The plants may grow to be as much as three feet tall, bearing elliptical to broadly lanceolate leaves. Mountain bluebells are a browse plant for elk and deer, and it is common to find matted areas where large animals have bedded down, sometimes to give birth to their young in the thick plant growth. The ciliate bluebell grows in all the Rocky Mountain states.

Idaho bluebell
Mertensia campanulata A. Nelson

The Idaho bluebell is found only in central Idaho. A tall meadow plant, it often grows in profusion in well-circumscribed areas as shown in the illustration above. The leaves, stems and flowers all have a pronounced frosted (glaucous) appearance; their smooth leaves lack prominent veins. Idaho bluebells flower in late spring and then, with the summer's heat, the plants disappear. For whatever reason, they may not reappear in the same location in subsequent years. The word *campanulata* means "bell-shaped," a term that could equally well be applied to the flowers of most mertensias. The common name "bluebell" is used for many other plants—some quite unrelated—emphasizing the importance of binomial scientific names.

Oregon bluebell
Mertensia bella Piper

The Oregon bluebell is the least common of the several species of mertensia shown here, for, while it occurs in Idaho, Montana, Oregon and California, it is considered to be an uncommon plant in all four states; it is found only in Clearwater and Idaho Counties in Idaho. It may be identified by its thin, green leaves. Their upper surface is covered with fine hairs, and the undersurface is smooth. The leaves' veins are more prominent than those of our other mertensias. The flowers are open, and the small lobes at the end are rounded. The plant is usually found in moist surroundings, growing to mid-elevations in our mountains.

Leafy bluebell
Mertensia oblongifolia
(Nutt.) G. Don

The leafy bluebell is one of our earliest spring wildflowers, blooming on sagebrush-covered slopes soon after, or even during, snowmelt. The species name, *oblongifolia*, describes the plant's wide leaves. All mertensias have five sepals enclosing five petals. These form a tube that flares more or less abruptly. Albino forms of this plant are occasionally seen. Eighteen species of *Mertensia* grow in North America; the leafy bluebell is restricted to our western states. German botanist Karl Heinrich Merten (1796–1830) collected the plant while on a Russian scientific expedition to Alaska in 1827. The genus *Mertensia* was named in honor of his father, Franz Karl Merten (1764–1831), also a botanist.

Columbia puccoon
Lithospermum ruderale Douglas ex Lehm.

The Columbia puccoon (members of this genus are also known as gromwells or stoneseeds) is a western plant ranging north to Alberta and British Columbia, south through the Rocky Mountain states to Colorado, Utah and Nevada and west to the three coastal states. It is a moderately tall plant with prominently ribbed leaves. The leaves and stems are coated with fine hairs giving the plants an overall grayish-green appearance. Its seeds (nutlets) are bony hard, explaining its scientific name, *Lithospermum* ("stone-seed," derived from the Latin). The species name, *ruderale*, also Latin, means "dump" or "waste-place," although the plant is no more commonly found in disturbed areas than are many other species of wildflowers.

Seeds of the European *Lithospermum officinale*, and possibly of ours as well, were used medicinally in times past to treat bladder stones in the belief that "like cures like." The roots of the Columbia puccoon contain a yellow dye used by Native Americans. The roots of the related eastern plant, *Lithospermum canescens*, contain a red dye. Captain John Smith wrote in 1612: "Pocones is a small roote that groweth in the mountaines, which being dryed and beate in powder turneth red." It was used by the Indians to paint their skin. Smith's observation may have been the origin of the term "redskin."

Common fiddleneck
Amsinckia menziesii
(Lehm.) A. Nelson & J. F. Macbr.

The common (also known as small-flowered) fiddleneck is a widely distributed native weed, growing to subalpine elevations. In common with many Boraginaceae, it has bristly stems, leaves and flower parts. The flowers bloom on an unrolling stem known botanically as a helicoid or scorpioid cyme, from which the name "fiddleneck" was derived. The plants sometimes grow in such numbers as to turn fields yellow. Livestock avoid the stiffly bristled plants, so the fiddleneck is classified as a troublesome weed. The plant was first described by German botanist Johann Georg Christian Lehmann (1792–1860), who gave it the generic name *Amsinckia* to honor a nineteenth century benefactor of the Hamburg Botanical Garden, William Amsinck. Lewis and Clark collected this species on the Columbia in 1806, although Archibald Menzies (1754–1842) surgeon and botanist with the Vancouver Expedition (1791–1795), had found it earlier, explaining its present species name.

---

*Menzies made the first ascent of Hawaii's Mauna Loa in 1794 accompanied by two companions. He measured the mountain's height quite accurately with an anaeroid barometer.

Nonnative Boraginaceae

Many wildflower guidebooks do not include nonnative plants, most of which are native to Eurasia. As a result, readers are at a loss as they attempt to identify them. Four nonnative Boraginaceae that grow wild in Idaho are described here:

The **German madwort, *Asperugo procumbens* L** (left) is a relatively non-aggressive field plant that grows to mid-elevations. Unfortunately, it has no value and is useless as a browse plant. In the past it was used (ineffectually, we presume) to treat rabies, hence the common name "madwort."

Common hound's-tongue, *Cynoglossum officinale* L. (center) grows to fairly high elevations and, increasingly, away from settled areas. It came to the New World centuries ago valued as a healing poultice and as an ornamental. The plant is now classified as a weed in many states. Hound's-tongue may grow to be two feet or more high. It is identified by its narrow gray-green, furry leaves and tightly clustered, dark red to purple flowers. The name "hound's-tongue" is the translation of the Greek name for the plant.

Field (or corn) gromwell, *Buglossoides* (*Lithospermum*)

arvense L. (right) is a thick-stemmed, hairy-leaved, straggly plant with small white or occasionally light blue flowers. It was used extensively in the past for its many supposed therapeutic benefits (diuretic, contraceptive, healing poultice, etc.). Its medicinal values are questionable, and the plant can cause liver damage both to humans and animals. As with the other introduced Boraginaceae, the field gromwell now grows in many localities throughout North America.

Viper's bugloss,* *Echium vulgare* L. (not shown) has several to many straight stems (spikes) that bear attractive dark red flowers. These turn deep blue as they age. The plant's hairy leaves and stems mark it immediately as a member of the borage family. Viper's bugloss was imported as an ornamental early on, and it is still a favorite garden plant. Unfortunately, it also thrives in the wild and is classified as a noxious weed in many states and provinces.

---

*The OED (*q.v.*) considers the etymologies of "viper"and "bugloss" at some length. Briefly, various plants were thought to neutralize snake venom. "Bugloss," in turn, came from a Greek word meaning "ox's tongue," presumably for the shape of the leaves.

Cabbage, or Mustard, Family (Brassicaceae)

The scientific family name Brassicaceae is derived from *brassica,* the Latin word for "cabbage." The family contains many food plants rich in vitamin C and sulfur compounds—the latter are responsible for the typical smell and taste of foods such as cabbage, brussel sprouts, broccoli, turnips, and mustard (all of which belong to the genus *Brassica*), water-cress (*Rorippa* spp.), radishes (*Raphanus* spp.) and others. While these are of economic importance, the family also contributes to our ornamental gardens: ornamental cabbages, wall-flowers (*Erysimum* spp.), rockcresses (*Arabis* spp.) and others. The flowers in this family all have four petals. These usually form a cross, explaining an older family name, Cruciferae, derived from the Latin *crus* ("cross") and *fero* ("I bear"). Members of the family are often referred to as "crucifers."

The flowers are frequently borne in clusters (racemes). Leaves are usually simple, alternate with each other on the stem, and typically lack a petiole (a leaf stem). Most of the Brassicaceae form seedpods. When these are long, they are known as siliques; when short, they are silicles. The leaves and other parts of the plants often have a radishy taste. Members are represented in our mountains by several genera. Many species, including some with showy flowers, are shown on the following pages. Most likely any small, four-petaled wildflower that you see blooming early in the spring will belong to the Mustard family, although further identification of the plant can be difficult as there are many similar species. Recent changes in classification have made identifying plants in this family even more difficult.

Cusick's rockcress
Boechera cusickii
(S. Watson) Al-Shebaz

Cusick's rockcress grows only in Idaho, Nevada, Washington and Oregon. Its flowers range from purple to nearly white. The plant may be identified by its crowded flower cluster, leafy stems and pendulous fruiting bodies or siliques (not present in the image). Its silvery-green color comes from fine hairs that cover the foliage. William Conklin Cusick (1842–1942) was an Oregon schoolteacher, rancher and botanist who collected and described many plants native to Oregon's mountains.

Arabis or Boechera?

Certain rockcresses, both in the New and Old Worlds, previously classified as *Arabis* have, on the basis of molecular studies, proved to be genetically different from others classified in the same genus. To accommodate these, a new genus, *Boechera* (the name honors Danish botanist Tyge W. Böcher, 1909–1983), was created. Most of the plants that fell into the new classification turned out to be native to North America. Change sometimes comes hard. The USDA and others continue (as of this writing) to use *Arabis* for all the rockcresses. We will follow the classifications we used previously, suggested by Dr. James L. Reveal, who helped us with our second edition.

Elegant rockcress, *Boechera sparsiflora* Nutt.

Elegant rockcress (also known as sicklepod rockcress) is an attractive foothill plant. It is distinguished by its long siliques—the upper ones pointing upward, the lower ones curving downward. The flowers are quite showy, especially as this plant may be quite tall with large flowers. These are typically a deep, intense purple, but may be light in color, even approaching white. Six varieties have been described; the one shown here, var. *sparsiflora*, is the variety commonly encountered in Idaho. As one variety or another, elegant rockcress grows in most western states as far south as Arizona and north to Yukon Territory in Canada.

Hoary rockcress, *Boechera puberula* (Nutt.) Dorn

The chances are that any pink to purple, spring-blooming, four-petaled plant found in our mountains will be a *Boechera*. This showy little plant's species name, *puberula,* implies that its gray-green leaves are covered with minute hairs. It grows in Idaho, in the three Pacific coastal states, Utah and Nevada. Many rockcresses prefer rocky ground, explaining their common name.

Holboell's rockcress
Boechera holboellii
(Hornem.) A. Löve & D. Löve

Holboell's rockcress is widely distributed. It is found in the West, Midwest, east to Quebec and north to Alaska and Greenland. As is common with plants having a wide range, many varieties are recognized; several occur in Idaho—the plant shown here is var. *secunda* (Howell) Dorn. Holboell's rockcress may be identified at the species level by its pendulous siliques. It is encountered fairly frequently, growing as high as the subalpine zone.

The species name, *holboellii*, honors an eminent Danish naturalist, Carl Peter Holboell (1795–1856), a man whose interests apparently varied greatly, for various species of birds, mollusks and fish also bear his name.

Wind River rockcress
Boechera williamsii (Rollins) Dorn
var. *saximontana* (Rollins) Dorn

This recently reclassified plant appears to be Williams, or Wind River, rockcress (identified as *Arabis microphylla* var. *saximontana* in the first edition of this book). It is an uncommon, tiny plant characterized by smooth-surfaced basal leaves, few-leaved stems and small clusters of attractive pink to purple, four-petaled flowers. Look for it shortly after snowmelt, close to treeline. Two varieties are recognized. One, var. *williamsii*, is found only in Wyoming where the species was first collected. The other, the plant shown above, grows in Idaho, Wyoming and Montana.

Nuttall's rockcress
Boechera nuttallii (Kuntze)
B. L. Rob.

Nuttall's rockcress is at home from the foothills to mid-montane elevations. A thin stem arises from a basal rosette of leaves, topped by a cluster of white to pinkish flowers that form upward-pointing siliques.

Englishman Thomas Nuttall (1786–1859), generally accepted as the greatest of our early botanists, spent years botanizing in the United States and western North America. His *Genera of North American Plants* appeared in 1818 and led to an appointment as curator of Harvard's Botanic Garden, a post he held for more than a decade. Then, following a productive overland trip to the Pacific coast, he returned to England and lived there for the remainder of his life.

Alpine bladderpod
Physaria reediana O'Kane & Al-Shehbaz

The alpine bladderpod (formerly *Lesquerella alpina* (Nutt.) S. Watson) blooms in the spring on dry subalpine ridges. The common name "bladderpod" is derived from the plant's round fruiting bodies (not obvious in the illustration). It is native to the northern Rocky Mountains and neighboring states, growing as far north as Alberta and south to Colorado.

Western bladderpod
Physaria occidentalis
(S. Watson) O'Kane & Al-Shehbaz

The western bladderpod (previously *Lesquerella occidentalis* S. Watson) is a small, more common plant than the alpine bladderpod shown on the left. It grows on gravelly ground from mid-elevations to subalpine slopes. Spoon-shaped leaves are unique to this species and serve to identify it, as does the centrifugal growth pattern of the clusters. It grows in Idaho, Nevada and Utah as well as in the mountains of the three Pacific coastal states.

Common twinpod (left)
Physaria didymocarpa (Hook.) A. Gray

The twinpod physaria is unusual as its fruit is split into two short, joined pods. Even when the plant is not fruiting, as with the plant in the illustration, it can be identified by its basal cluster of furry, gray-green, ovoid leaves and its four-petaled bright yellow flowers. This little plant is found at higher elevations in our mountains, growing on dry, gravelly ground. It grows in Idaho, Wyoming, Montana, North Dakota and Alberta (and rarely in Washington and British Columbia). The name *Physaria* is from a Greek word meaning "bladder" or "bellows" for the shape of the silicle (fruiting body). The species name, *didymocarpa*, is also from the Greek and means "twin fruit." Three varieties are recognized; the one shown here, var. *didymocarpa*, grows in Idaho.

Few-seeded draba (left)
Draba oligosperma Hook.
The species name, *oligosperma*, means "few seeds." The few-seeded draba grows in all but the most southern of our western mountain states and Canadian provinces, from lower elevations to subalpine slopes. While its flowers are similar to other plants shown on this page, its narrow leaves do not form mats.

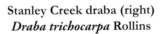

Globe-fruited draba (below)
Draba sphaerocarpa
J. F. Macbr. & Payson
Some drabas are localized to a single mountain range; this cushion-forming plant, for example, grows only on exposed alpine and subalpine ridges of the Sawtooth Mountains. Its common and species names both refer to its rounded fruit. While many draba species are similar, tentative identification is often possible based on plant location and morphology. Definitive identification often depends on technical differences among the fruits of various species.

Draba spp.
Made up of about 350 species, the drabas are the largest genus in the Mustard family. The name, from Greek antiquity, was used for a now unknown crucifer. "Whitlow grass" is a common name because draba poultices were used in the past to treat whitlows, or "run-arounds"—infections that form at the base of fingernails. At lower altitudes, drabas tend to be long-stemmed solitary plants with plain white, pink or yellow flowers. At higher elevations, they form compact clumps on rocky slopes, usually flowering at the end of snowmelt.

Payson's draba (above)
Draba paysonii J. F. Macbr.
Payson's draba is found only at high elevations in the Sierras and in the northern Rocky Mountains of the United States and Canada. The plant shown here was photographed well above treeline on Mount Borah, Idaho's highest mountain. Its tiny, hairy, clustered leaves grow at ground level and remain well hidden when the plants are in flower. Two varieties are recognized; ours is var. *treleasii* (O. E. Schulz) C. L. Hitchc.

Stanley Creek draba (right)
Draba trichocarpa Rollins
The Stanley Creek draba has the most restricted range of any of the very similar alpine/subalpine plants shown on this page, for it is found only near Stanley, Idaho. Because this species was described relatively recently, it is often not listed in regional plant guides. Its four-petaled flowers are small and remain partially closed, and its tiny leaves form tight clusters. The species name, *trichocarpa*, means "hairy fruit."

Lance-leaf draba
Draba cana Rydb.

The lance-leaf draba (also known as the cushion whitlow-grass) blooms later in the spring than do those shown on the previous page, favoring high open slopes and rocky crevices. This plant's classification has been a source of confusion in the past. It has variously been classified as *Draba lanceolata* or as a variety of *Draba breweri*. The classification above is correct. The plant is found throughout the Rocky Mountains to Alaska, east across the continent to several northern states and south to California, Nevada, New Mexico and Utah. It also grows in Greenland and Eurasia.

Alpine smelowskia
Smelowskia americana (Regel & Herder) Rydb.

The alpine smelowskia (until recently classified as *Smelowskia calycina*) is a somewhat protean plant with hairy, variably pinnate basal leaves, three-lobed terminal leaflets, ovoid narrow-based petals and quite prominent anthers and styles. Its fruit is ovoid with a protuberant tip—a few are seen in the illustration. Alpine smelowskias prefer rocky alpine and subalpine terrain. It grows in the central and northern Rocky Mountains, north to Alaska. The name honors Timotheus Smelowski (1770–1815), a Russian botanist. "False candytuft" has been proposed as an alternate common name for smelowskias.

Idaho candytuft
Noccaea fendleri (A. Gray) Holub var. *idahoense* (Payson) F. K. Mey.

Two native thlaspis, formerly varieties of *Thlaspi fendleri*, have recently been reclassified into separate species: *Noccaea montana* and the plant shown here (until recently classified as *Thlaspi idahoense* Payson). The two are distinguishable chiefly by the configuration of their leaves. The leaves of the former are rounder and have well-developed petioles (leaf stems), whereas the leaves of the Idaho candytuft are more lanceolate and taper gradually to attach directly to the main stem, as seen in the illustration. Our plant grows in the mountains of central Idaho, whereas the alpine species is found in all our western states. The genus *Noccaea* honors Domenico Nocca (1758–1841), an Italian clergyman and botanist who was the director of the botanical garden at the University of Pavia.

Western tansy mustard
Descurainia pinnata (Walter) Britton

Descurainia is a confusing genus. All its members are weeds. A few are imports, but most are native plants. Ours appears to be a variety of *Descurainia pinnata*, although we will not attempt to classify it further as no less than ten varieties are recognized, based on relatively minor differences in morphology. The plant shown here is very common, growing to mid-elevations in our mountains. The species, in one variety or another, occurs throughout the United States. It is characterized by clusters of small four-petaled yellow flowers, erect siliques (just forming here) and pinnate leaves—whence the species name *pinnata*. "Pinnate tansy mustard" is another, more appropriate, common name.

Young plants are similar in appearance to those of the flixweed (*Descurainia sophia* (L.) Webb ex Prantl, not shown), a troublesome import that has spread all across North America. As the flixweed matures, it develops frizzy, bi- or tripinnate leaves, distinguishing it from native species. The genus *Descurainia* was named for French botanist François Descurain (1658–1740).

Western wallflower
Erysimum asperum (Nutt.) DC. var. *elatum* (Nutt.) Torr.*

Half a dozen varieties of western wallflower are recognized. Ours is found in Idaho, Utah and Nevada. It grows on rocky slopes—often on talus—at mid- to subalpine elevations. Its growth habit and its large, showy, yellow-orange flowerheads identify this attractive plant. After the petals fall away, its flowers form siliques, slim seed-containing pods. The generic name, *Erysimum*, was used by the Greeks for a related plant. Lewis and Clark collected the western wallflower while camped near present-day Kamiah, Idaho, on June 1, 1806.

The species name, *asperum*, means "rough," but we don't know why it was used for this plant, whereas the varietal name, *elatum* (tall), is quite suitable.

---

*This plant is a member of a group of very similar plants about which there is taxonomic debate; we have chosen to use the classification above. The plant is also known as Pursh's wallflower and then may be classified as *Erysimum capitatum* (Douglas ex Hook.) Greene var. *purshii* (Durand) Rollins.

American wintercress (or yellow rocket)
Barbarea orthoceras Ledeb.

The American wintercress shown here (and misidentified as a worm-seed mustard in the second edition of this book) is a wetland plant that grows to high elevations in our mountains and elsewhere. It is widely distributed in North America, where it is found in almost every state and every Canadian province and territory, and east to Greenland and Eurasia as well. The plant is characterized by tightly clustered, four-petaled yellow flowers and radish-like leaves with basal lobulations. As its common name suggests, the plant is edible. We have not tried it, but suspect that it would serve well as a salad green.

Watercress
Rorippa nasturtium-aquaticum (L.) Hayek

Watercress (also classified as *Nasturtium officinale* R. Br.) is said to be an exotic plant of Eurasian origin. It grows in almost every state and in most of the Canadian provinces; indeed it, like the wintercress shown above, is so widely distributed that it is hard to believe it was not here when Europeans arrived. It is easily identified, for it dwells in slow-moving water, blooming in late summer. The white flowers and the elongated fruit (siliques), shown in the illustration, are typical of the Brassicaceae family as a whole. Watercress leaves are crisp with a peppery taste, more pronounced in wild than in cultivated plants. (Warning: the water in which the cress grows may be contaminated with the intestinal protozoan parasite *Giardia lamblia*.) The name *Rorippa* is said to be derived from the Anglo-Saxon word for the plant. The Latin species name, *nasturtium-aquaticum,* means "water nasturtium," from the similar peppery taste of watercress and the South American nasturtium, a member of the genus *Tropaeolum*.

Blue mustard (left)
Chorispora tenella (Pall.) DC.

This common weed—it is sometimes known as the cross-flower—is easily identified, for there is no other crucifer in our area with similar small, four-petaled, light purple flowers. It seems to be spreading rapidly in Idaho. The name *Chorispora,* derived from two Greek words, means "separate seed" for the configuration of the seeds in the plant's silique. If cows browse on this plant, their milk will have an unpleasant taste.

Whitetop (above)
Cardaria draba (L.) Desv.

The whitetop (also known as heartpod hoary cress) has compact, flat-topped flower clusters that bloom from the outside in. It is further identified by its rather succulent serrated leaves and heart-shaped silicles. It blooms from early spring through the summer, sometimes forming immense patches on disturbed ground.

Spring whitlow-grass (below)
Draba verna L.

As with the other weeds shown here, the spring whitlow-grass has spread throughout much of the Northern Hemisphere. It is easily identified by its clustered bright white flowers. The flowers are unusual for they have two pairs of opposed, deeply cleft petals—an identifying feature.

Field pennycress (left)
Thlaspi arvense L.

Field pennycress grows on disturbed dry ground. It is so common a weed that most will recognize it at first glance. Its common name, "pennycress," is derived from its disc-shaped silicles. A similar weed, the shepherd's purse, *Capsella bursa-pastoris* (L.) Medik., (not shown here), takes its name from its triangular fruiting bodies.

Cactus Family (Cactaceae)

The Swedish botanist Carl Linnaeus (1707–1778) published his *Species Plantarum,* the basis for our modern system of binomial plant classification, in 1753. Challenged to find names for the many plants that he listed, he often used ones mentioned by ancient writers. He knew of American cacti, and he bestowed on them the Greek name *kaktos,* believing that ours were related to a relative of the artichoke that the Greeks knew by that name. Later, the cacti were classified as a separate family, one that contains about 1,700 species. Cactaceae are New World plants (a cactus [*Rhipsalis* sp.] from Madagascar and southern Africa may have been introduced), ranging throughout the Americas from British Columbia to the southern ends of Chile and Argentina. The cacti have adapted superbly to the extremes of heat and dryness encountered in desert environments. In most, the leaves have been replaced with spines, and photosynthesis goes on in the plants' fleshy stems. Some bear edible fruit. The stems take many forms: round, flattened, cylindrical, etc., and are often ribbed. The flowers are usually solitary, often showy, and have many sepals, petals and stamens. Four species of cactus are found in Idaho, growing to montane elevations. The ones shown here are the cacti most commonly seen.

Spiny prickly pear
***Opuntia polyacantha* Haw.**

The spiny cactus favors desert-like surroundings, where it grows to fairly high elevations. The flat stem-joints are covered with many spines, the meaning of *polyacantha*. The plant's attractive flowers range from light yellow to bright red. *Opuntia polyacantha* is found throughout the western United States and Canada. One might guess that the genus name, *Opuntia,* was derived from a Native American word, but no, it was a word taken over by Linnaeus, that of a now unknown plant that grew near the city of Opus in ancient Greece.

Brittle cactus
Opuntia fragilis (Nutt.) Haw.

The small, ground-hugging brittle cactus grows as high as the montane zone, nestling against exposed rocks that hold the day's heat. At these higher elevations the plants—usually green—take on a reddish hue, as shown in the inset. The species name, *fragilis*, reflects the ease with which segments break off the main stem and attach themselves to passersby. Showy flowers last only a day or so—a photographer must be vigilant to catch them—followed by spiny red fruit. This fragile cactus grows throughout the western part of the United States (except in Nevada and California) and Canada.

Mountain cactus
Pediocactus simpsonii
(Engelmann) Britton & Rose

The mountain cactus is said to be the highest growing of our native cacti (although in Idaho *Opuntia fragilis* is a contender). It prefers mid-elevation mountain valleys and grows in all the Rocky Mountain states, including Oregon and South Dakota. The plant is notably variable (it was incorrectly identified as a nipple cactus, *Escobaria missouriensis* [Sweet] D. R. Hunt, in our last edition). Identify it by its shape and its prominent bundles of spines. The attractive flower may be more than an inch across with many "petals" (actually tepals). These may be colored, although ours are usually white. Attempts at cultivation are mostly unsuccessful.

Honeysuckle Family (Caprifoliaceae)

The Honeysuckle family includes five genera and about 210 species. Many are native to the Americas and to Eurasia, growing in the vicinity of the Mediterranean. Most members of this family are hardy shrubs and vines. These include various honeysuckles and snowberries, commercially important as ornamental garden plants.

Although we will treat all the plants shown here as Caprifoliaceae, it should be noted that five genera and approximately 245 species were recently assigned to the Moschatel (Adoxaceae) family. Members of this family, including elderberries and viburnums, mostly grow in temperate climates as far north as the arctic regions of the Northern Hemisphere. Taken together, as was traditionally done until the 1990s, the members of these two families have flowers with five petals joined to form a basal tube, and five sepals. The leaves are mostly opposite and lack petioles (stemlets). Twinned flowers and fruit are common, as with garden honeysuckles (*Lonicera* spp.).

The Honeysuckle family's scientific name, Caprifoliaceae, was derived from an old word, "caprifoil," used for the European honeysuckle *Lonicera periclymenum.* Caprifoil, in turn, came from the Latin *caprifolium,* derived from the Latin *caper* for "goat" and *folium* for "leaf." There is some suggestion that the name for the plant may have originated on the island of Capri, where goats were common.

Trumpet honeysuckle
Lonicera ciliosa (Pursh) Poir. ex DC.

The attractive, orange-flowered trumpet honeysuckle vine is often seen growing along roadsides and trails as high as montane elevations. It is native to the four northwestern states, British Columbia and the northern part of California. The plants are easily identified by their clustered, long, bright orange, trumpet-shaped flowers and by a pair of conjoined opposing leaves just below the flower cluster through which the stem passes (a leaf arrangement sometimes encountered in other plants in the honeysuckle family). The trumpet honeysuckle was unknown to science until Lewis and Clark returned to the United States with a dried specimen that they had collected on June 5, 1806, while camped on the Clearwater River near today's Kamiah, Idaho—a location close to where this plant was photographed.

Twinberry (above, both images)
***Lonicera involucrata* (Richardson) Banks ex Spreng.**

The twinberry is a shrub found along watercourses as high as the subalpine zone. Inconspicuous twinned yellow flowers appear in the spring, and by midsummer paired, inedible, blue-black berries appear. These form above two dull red leaves, or bracts, together known as an involucre, from which the species name was derived. Under favorable conditions this turns a bright waxy scarlet as in our illustration. The genus *Lonicera* takes its name from Adam Lonitzer (1528–1586), a German botanist. Meriwether Lewis collected this plant near the Continental Divide in Montana on July 7, 1806, not knowing that the plant had previously been described. Twinberries are found in all the western states and most Canadian provinces.

Utah honeysuckle (below, both images)
***Lonicera utahensis* S. Watson**

The Utah honeysuckle grows in the mountains of Idaho from mid-elevation nearly to treeline. Its small flowers are neatly paired and their ovaries mature into two red berries that sometimes fuse into hourglass-shaped fruit. An attractive shrub, it is often used in ornamental landscaping. Lewis and Clark collected a specimen of the Utah honeysuckle in fruit, most likely while ascending the North Fork of the Salmon River, on September 2, 1805. The Utah honeysuckle grows in the northwestern states, Utah, Colorado, New Mexico, Arizona and the Canadian provinces of British Columbia and Alberta.

Caprifoliaceae

Mountain snowberry (above)
Symphoricarpos oreophilus A. Gray

Common snowberry (below)
Symphoricarpos albus (L.) S. F. Blake

The mountain snowberry replaces the common snowberry at higher altitudes. It is found all along our hiking trails, and is easily recognized by its distinctive soft, blue-green, oval leaves. In late spring, the shrubs bear the small, white, tubular flowers shown above. They are usually borne in pairs, with petals that flare out a bit at the end. The generic name, *Symphoricarpos,* is derived from three Greek words that mean "fruit that is borne together" in pairs. The species name, *oreophilus,* also is from the Greek and means "mountain lover." This plant's berries, shown above, are smaller and oval-shaped compared to those of the common snowberry.

The common snowberry is found throughout our western mountains and in all the northern states and Canadian provinces, growing at mid-elevations. It is commonly encountered along our trails and roadsides. Our plant is var. *laevigatus* S. F. Blake; the varietal name means "smooth." It was unknown to science until collected by the Lewis and Clark Expedition, probably somewhere along the Missouri River on an unknown date. This snowberry has been used as an ornamental shrub almost from the time the explorers' specimen (dried fruit) grew out from their seeds, planted in Philadelphia. It remains a popular garden plant today. The mushy white berries, shown below, have no food value.

Western blue elder (left)
Sambucus cerulea Raf.
Western black elder (right)
Sambucus racemosa L.
var. *melanocarpa* (A. Gray) McMinn

The two elders shown here grow as shrubs or small trees. The plants are quite similar. Both are native to most of our western states and Canadian provinces. Elders are often seen growing along streambanks and in permanently moist areas as high as the subalpine zone. Their compound leaves are odd pinnate; each has seven toothed leaflets. As the illustrations suggest, the flower clusters of the two species differ slightly. The blue elder's flower cluster is flat-topped, whereas the black elder has cone-shaped clusters. The color of their fruit is reflected in their names; the species name, *cerulea,* means blue, and the varietal name of the western black elder, *melanocarpa,* means "black fruit." The genus name, *Sambucus,* is derived from the Latin name for an Old World elder; the species name, *racemosa,* was derived from "raceme," a botanical term for a flower cluster that blooms from the bottom upward. Elderberries ripen in late summer and are often used to make jellies and wine. The berries do not last long in the wild, for they are a favorite of birds and other animals.

Pink Family (Caryophyllaceae)

The Pink family is moderately large, consisting of ninety-three genera and 2,400 species scattered throughout the north temperate zone; many are native to the northern Mediterranean countries. The family is well represented in our Northwest. While many of the European species are colorful, most of ours are unprepossessing small white flowers. Typically, plants in the Pink family have narrow, opposed leaves that originate from swollen nodes along the stem. The flowers are usually five-petaled, and the ends of the petals are often notched or fringed. In some species the sepals coalesce to form a swollen, tube-like involucrum. Ornamental plants and cut flowers, especially species of *Dianthus,* have considerable economic importance. A few of our Caryophyllaceae are weeds, including several garden ornamentals originally native to Europe that now grow wild in the western United States. These include bouncingbet (*Saponaria officinalis*), babysbreath (*Gypsophila paniculata*), chickweed (*Stellaria media*) and several others.

The scientific name Caryophyllaceae is derived from the Greek word for the clove pink (the original carnation) that has a clove-like odor; in fact, the Greek word for cloves (*garyophalla*) and for the plant (*garyophylla*) are almost identical. In forming the plant name, the Greek words *karyon* for "nut" and *phyllon* for "leaf" were combined to describe the dried buds of the clove tree (*Syzygium aromaticum*).

Sticky chickweed
Pseudostellaria jamesiana (Torr.)
W. A. Weber & R. L. Hartm.

The sticky chickweed (formerly *Stellaria jamesiana*) bears five-petaled flowers that bloom from late May into June, or later depending on elevation. It prefers shaded places. The plant is native to all our western states including Texas. It may be identified by its long leaves with prominent central ribs and by the flower's dainty notched petals. Edwin James (1797–1861), whom the species honors, was a surgeon-naturalist with the Long Expedition that explored Colorado's Rocky Mountains in 1820. The name "chickweed" comes from related European species whose seeds apparently are a favorite food of poultry. The common name "sticky starwort" has been suggested for this plant to avoid confusion with other chickweeds.

Long-stalk starwort	**Nodding mouse-ear chickweed**	**Field chickweed**
Stellaria longipes **Goldie**	***Cerastium nutans*** **Raf.**	***Cerastium arvense*** **L.**
		var. ***strictum*** **(L.) Ugborogho**

The long-stalk stellaria, or starwort, is a common native plant that grows at higher elevations. It may be identified by its tall (*longipes*), thin, four-sided stalk, long pointed leaves—two arising at each node—and notched petals. The plant prefers moist situations, appearing in fields following snowmelt and on streambanks. It is widely distributed throughout North America (including Greenland and other arctic islands) and in all but our southern central and eastern states.

To a casual observer there seems to be little difference between plants in the genus *Stellaria* and those in *Cerastium*. In fact, the plants are quite similar; the differences are mostly technical based on the form of the seed capsules. This plant's species name, *nutans*, means "nodding," although this is seen chiefly in budding plants rather than in mature flowers. The plants tend to grow in loose clumps along streambanks where they spread by underground stems (rhizomes). The similarities of this family's many small plants, one to another, all with family characteristics of opposing leaves arising in nodes along the stem and notched petals, make identification below the family level diffficult. Add many similar nonnative species to the mix and one is left with a confusing group of little white wildflowers.

The field chickweed is a common, widely distributed wildflower found at all elevations throughout North America, in South America and in Eurasia. It is characterized by deeply notched petals and narrow, opposing, gray-green leaves. The name *Cerastium* was derived from a Greek word, *kerastos*—meaning "horn"—for the shape of its seed capsules. Given the plant's wide distribution, it is not surprising that there is considerable variation within the species. At least six varieties are recognized (*strictum* means "erect" or "straight," apparently referring to the long stem). The species name, *arvense*, from the Latin, implies an open meadow, reflecting the plant's growth preference.

Parry's silene
Silene parryi (S. Watson)
C. L. Hitchc. & Maguire

Parry's silene blooms from early summer on, according to the altitude; it ranges as high as treeline. It is a long-stemmed plant with a cluster of basal lanceolate leaves. It may be further identified by notched petals that barely protrude from the end of its square-based, bottle-shaped calyx tube formed by five joined and pointed sepals. As the flowers age, ten longitudinal green stripes on the calyces turn deep purple. The plant is found in British Columbia and Alberta, throughout the Northwest and south to Idaho, Oregon and Wyoming. Charles Christopher Parry (1823–1890), for whom this species was named, was an English-born American botanist. Several other western plants bear his name.

Scouler's silene
Silene scouleri **Hook.**

Scouler's silene is found in most of the western mountain states and Canadian provinces. At least three, and usually more, pairs of broadly lanceolate, opposing leaves become increasingly smaller and farther apart as they ascend the stem. Higher up there are two or more tight clusters of long-tubed flowers with hairy, sticky calyces. Scouler's silene is a rather common species, found mostly at lower altitudes. The variety shown here is var. *concolor* (Greene) C. L. Hitchc. & Maguire. This plant's species name honors John Scouler (1804–1871), a ship's surgeon who came to America in 1825 with David Douglas. Douglas remained, and Scouler returned with his ship to England, having found several new plants during his brief stay in the Northwest.

Oregon silene
Silene oregana S. Watson

The Oregon silene grows in every state contiguous to Idaho as well as in California. A tall plant, it is characterized by a basal cluster of markedly furry leaves as shown in the photograph. A long, thin stem is topped by one to several flowers. Swollen calyceal tubes end in white, five-petaled flowers. Each petal is deeply cleft into four lobes. These are often frizzy-ended. The Oregon silene is a perennial that grows quite high in our mountains on exposed ridges and in open groves of firs and ponderosa pines.

Menzies' silene
***Silene menziesii* Hook.**

Menzies' silene has open clusters of small flowers, each with five white petals. The petals are split into two lobes, each of which has an inner tab-like appendage at its base. This is a common feature of flowers of this genus, although the appendages are not always easy to see. All silenes have enlarged calyceal tubes, although the tubes are not as prominent in this species as in others. This little plant favors open woods, where it may grow to fairly high elevations. It is found throughout the West, south to Arizona and New Mexico and north to Alaska. Its name honors Archibald Menzies (1754–1842), surgeon-naturalist with the Vancouver Expedition, who collected many new-to-science plants during the exploration of our Northwest.

Uinta sandwort
***Eremogone kingii* (S. Watson) Ikonn.**
var. *glabrescens* (S. Watson) Dorn

The Uinta sandwort is a low, spreading plant that grows at all elevations on rocky slopes and gravelly ridges, blooming from late spring well into the summer. Brown anthers borne on thin filaments overlie each petal and give the five-petaled flowers a spotted appearance. Until recently, this and similar plants were included in genus *Arenaria*, a term derived from the Latin word *arena* meaning "sand," reflecting the plants' preferred habitat. This plant's species name honors geologist Clarence King (1842–1901), who surveyed the Rocky Mountains and Great Basin in the 1860s. *Eremogone*, from two Greek words, means "solitary seed."

Nuttall's sandwort
***Minuartia nuttallii* (Pax) Briq.**

Nuttall's sandwort is found in all the northwestern states and Canadian provinces. Until recently classified as an *Arenaria*, it is now known by an earlier name that has priority. The genus name, *Minuartia*, honors Spanish botanist Juan Minuart (1693–1768). The common name "brittle stitchwort" has been suggested, reflecting the plant's tendency to break at nodal intervals along the stem.

| Common soapwort | Maiden pink | Bladder campion or catchfly |
| *Saponaria officinalis* L | *Dianthus deltoides* L. | *Silene vulgaris* (Moench) Garcke |

Common soapwort
***Saponaria officinalis* L**

This common import goes by many names; it is most commonly known as bouncing bet or sweet william. Long a favorite garden ornamental, it also does well in the wild where it is usually found growing in moist places. It is now native to all the lower forty-eight states. It does have some value as a useful herb, for as its name suggests, it contains saponin, a mild cleansing agent sometimes used instead of soap.

Maiden pink
***Dianthus deltoides* L.**

The maiden pink is a far less invasive garden escapee than is the soapwort shown on the left. As the image suggests, it is closely related to the carnation, *Dianthus catyophyllus*. The pink grows wild in many of our states and in several Canadian provinces. It is usually encountered growing in moist places.

Bladder campion or catchfly
***Silene vulgaris* (Moench) Garcke**

The bladder campion is a well-recognized weed found in all but a few southern states, in Alaska, and in most Canadian provinces. It is hard to see why the plant is so widespread for, despite its unusual appearance, it is not an ornamental and—in North America at least—is not recognized as having either food or medicinal value.*

---

*This is not so in northern Mediterranean countries—Spain and Greece particularly—where the young plants are collected and eaten as a delicacy.

Three Nonnative Caryophyllaceae

Many, if not most, of our native American plant families have been augmented over the years by Eurasian and Mediterranean imports. Most exotic members of the Pink family probably are ones that settlers brought with them for their ornamental and supposed medicinal value. The three plants shown here were all photographed growing wild in Idaho.

Goosefoot Family (Chenopodiaceae)

The Goosefoot family is moderately large, made up of ninety-seven genera and 1,300 species worldwide. It is well represented in the United States.* Both the common and scientific names were derived from the goosefoot plant, *Chenopodium album* (*chenopodium,* from the Greek, means "goose's foot"), a plant that was formerly used in Europe as a potherb known as "lambs-quarters" or "fat-hen." The same plant also occurs throughout North America, where little thought is given to it as a possible food source.

The family is chiefly notable for species that are important as food plants. These include spinach (*Spinacea oleracea*), which has largely displaced lambs-quarters as a green; Swiss chard and beets (*Beta vulgaris*) whose cultivars include both edible and sugar beets; blites (formerly included in the genus *Blitum,* but now classified as species of *Chenopodium*), one of which is illustrated below; and the South American cereal grains quinoa (*Chenopodium quinoa*) and amaranth (*Amaranthus* spp).

Some Chenopodiaceae are troublesome weeds, including various xerophytes and halophytes (respectively, plants that thrive in dry or salty environments).

The family is diverse and includes many weedy members. One or another species of weedy chenopods seem able to spring up almost overnight on newly disturbed ground. Many members of the family, including spinach, contain oxalic acid, a substance that is acutely poisonous when ingested in large amounts. Chronic ingestion of oxalic acid may also contribute to the gradual formation of oxalate-containing kidney stones.

---

*There is taxonomic debate, based on molecular studies, as to whether the Goosefoot family, Chenopodiaceae, should be considered a subfamily of the Amaranth family, Amaranthaceae. We have considered it to be its own family here.

Strawberry blite
Chenopodium capitatum (L.) Ambrosi

One may be surprised and curious on first seeing this chenopod, for the fruit looks for all the world like a ripe raspberry. Although not a wildflower per se, we have included it here because strawberry blite is not well known, and anyone who hikes in our mountains will sooner or later encounter the plant—one that is notable for its fruit, rather than for its unimpressive flowers.

Both the leaves of young plants and the ripe fruit are edible. The plant grows to fairly high elevations in our mountains. It is widely distributed and is found in all but the southern Great Plains states and the Deep South. It is also native to Eurasia. Its species name, *capitatum,* means "growing in a head," although why that was used for this plant is uncertain. The word "blite" (sometimes written, incorrectly, as "blight") is a term that goes back to ancient Greece (*bliton*) where it likely was used for this and other edible chenopods. Despite their succulent appearance, the berries are tasteless.

Cleome Family (Cleomaceae)

The Cleome (pronounced klee-OH-mee) family is small, consisting of eleven genera and about 300 species. Most grow in the temperate zones of both the Old and the New World; a few are found in Idaho, including the two plants shown below. The family is closely related to both the Mustard family (Brassicaceae) and the Caper family (Capparaceae) in which they were formerly classified. The relationship of the three families is so close, in fact, that some botanists consider all three to be Brassicaceae. The family has little economic importance other than the use of several members—the cleomes especially—as ornamental garden plants.

Rocky Mountain beeplant, *Cleome serrulata* Pursh

Originally found only in the West and Midwest, the Rocky Mountain beeplant is now found growing wild in the Northeast as an introduced plant. In Idaho, it grows as high as montane foothills. Beeplants are tall—up to four feet high—their stems surrounded by short, narrow, three-lobed leaves. The flowers are attractive, so cleomes are often grown as ornamentals. Lewis and Clark collected the Rocky Mountain beeplant in South Dakota in August 1804; the plant was then new to science.

Yellow beeplant, *Cleome lutea* Hook.

A closely related plant, the yellow beeplant has remained west of the Mississippi. A plateau plant, it prefers open terrain. Bees are attracted to cleome flowers, adding to their appeal. We are including this plant here; supposedly it occurs in Idaho as a montane plant, although—unlike the Rocky Mountain beeplant—we have not seen it growing that high. The young plants and the seeds were eaten by Native Americans. It is said that thorough boiling is necessary, for the plants have an unpleasant odor.

Dogwood Family (Cornaceae)

The Dogwood family is a small one, made up of one genus and sixty species. It is represented in our Northwest by three species. Two of these are flowering trees, and one is a small non-woody plant. All are attractive, well suited for use in gardens and ornamental landscaping—the family's chief economic importance. Their blooms are characterized by large, petal-like white bracts, usually four, but sometimes as many as seven, and by centrally clustered small and inconspicuous flowers.

The origin of the generic name, *Cornus*—from which both the family name, Cornaceae, and the alternate common name, "cornel" (used more often in Europe), were derived—is in doubt. The word means "horn" in Latin; possibly referring the trees' hard wood. The origin of "dogwood" is said to have been from skewers ("dogs") that butchers in the distant past made from the hard wood of the European dogwood, *Cornus sanguinea*.* The *Oxford English Dictionary* lists "dogwood" as being used first in 1676, in a citation describing "a fine Flower-bearing-Tree" that grew in Virginia, referring to the flowering dogwood, *Cornus florida,* which must have been introduced into England by that date.

---

*(I have been unable to find documented use of the word "dog") to mean a "skewer" *per se*. Nevertheless, the OED (q.v.) gives several examples of "dogs" used as sharp tools to fix or fasten objects.

Bunchberry, *Cornus canadensis* L.

The bunchberry is a ground-dwelling, non-woody plant, widely distributed across northern North America, west to Asia and east to Greenland. It is found in all our northwestern and northern tier states as far south as Colorado and (rarely) in New Mexico. It most commonly grows in shaded, moist forest surroundings. The plants have a rosette of five or six deep green leaves. In early summer four large white bracts, and later, bright red berries make identification easy. In common with other dogwoods, the bunchberry's actual flowers—seen in the center of the photograph above—are tiny. Given the requisite shaded situation, bunchberries do well in ornamental gardens.

Meriwether Lewis might have seen bunchberries flowering while the expedition was outward bound in the spring of 1805, and then seen fruiting plants during the east-to-west September crossing of the Bitterroot and Clearwater ranges later that year. We know that he also saw the plants flowering in Idaho in the spring of the following year, for he collected a specimen on June 16, 1806, as the homeward-bound expedition made its first attempt to ascend the Lolo Trail.

Red-osier dogwood, *Cornus sericea* L. (left and above)
The red-osier dogwood is a red-barked shrub or small tree that grows throughout Canada and in all but our southern states. The ovaries of its clustered, small, four-petaled flowers mature into white berries. "Osier," Latin for "willow," is a word used for pliant branches suitable for basket-making. The Latin species name, *sericea,* means "silky" for the fine hair on the leaves. Lewis and Clark wrote in the winter of 1804–05, while at Fort Mandan, that Indians smoked the bark of the red-osier dogwood mixed with tobacco, to stretch the tobacco supply. Both plants on this page are ones that are favored by landscape gardeners.

Western flowering-dogwood (right)
***Cornus nuttallii* Audubon**
The western dogwood grows mostly west of the Cascade range, but some are also found along Idaho's Clearwater River. Its resemblance to the eastern flowering-dogwood, *Cornus florida* L., may explain why Meriwether Lewis did not collect it in the spring of 1806. Later, Thomas Nuttall recognized that it was a new species. John James Audubon (1785–1851) included the western flowering-dogwood in one of his illustrations by way of thanking Nuttall for the use of ornithological material. This was the first published description of the tree, so Audubon is credited with establishing the scientific name.

NUTTALL'S DOG-WOOD
CORNUS NUTTALLII *Audubon*
This very beautiful tree, which was discovered by Mr. **NUTTALL** on the Columbia river, attains a height of fifty feet or more, and is characterized by its smooth reddish-brown bark; large, ovate, acuminate leaves, and conspicuous flowers, with six obovate, acute, involucral bracteas, which are rose-coloured at the base, white towards the end, veined and reticulated with light purple. The berries are oblong, and of a bright carmine.*

The Band-Tailed Dove or Pigeon
Columba fasciata Say

Image and the first published description of *Cornus nuttallii* from:
The Birds of America from Drawings Made in the United States and Their Territories
by John James Audubon *F.R.SS. L.E.E.*
New York, 1842, Vol. IV, page 312

*Although Audubon described six "involucral bracteas," the number varies from four to seven.

Stonecrop Family (Crassulaceae)

The Stonecrop family consists of thirty-three genera and approximately 1,500 species. Crassulaceae are found on all continents with the exception of Australia and Antarctica. Most of its members are thick-leaved succulents. The family takes its scientific name from the word *crassula*, an older generic term for the sedums and now the name of another genus in the family. The word *crassula* probably was derived from the Latin *crassus* meaning "thick" for the succulent leaves found throughout the family. *Sedum*, the genus to which the two plants shown below belong, is the largest genus in the family. The family's only economic importance is that many species—for example, the jade plant (*Crassula ovata*) from East Africa, the flowering red kalanchoe (*Kalanchoe blossfeldiana*) from Madagascar, and hen and chickens plants (*Jovibarba* spp.)—are popular ornamentals. The thick leaves of these and other species give many of the Crassulaceae an appealing, exotic look. It also means that they are easy to care for, given their capacity to store water.

Lanceleaf stonecrop, *Sedum lanceolatum* Torr.

The two stonecrops shown here are closely related plants. While both plants grow on rocky soil up to alpine elevations, they each have distinguishing characteristics. For example, the leaves of the lanceleaf stonecrop (shown above) remain close to the stem and are gradually shed as the plant matures. When the mature plant's seed capsules (follicles) form, they stand upright. The flowers of both plants are bright yellow, but the lanceleaf stonecrop's petals are shorter and wider. The plant is quite at home rooted in declivities, explaining the common name "stonecrop."

Wormleaf stonecrop, *Sedum stenopetalum* Pursh

The wormleaf stonecrop differs from the related lanceleaf sedum by its longer leaves (explaining "wormleaf") that stand out from the plant's stem (as do the follicles), where they are retained as the plant matures. The wormleaf stonecrop is found only in the northwestern states, California and the two western Canadian provinces, whereas the lanceleaf plant grows further south to Arizona and New Mexico, north to Alaska and east to the Great Plains. Meriwether Lewis collected both of these plants on the same day (July 1, 1806), while camped at Traveler's Rest near today's Missoula, Montana.

Heath Family (Ericaceae)

Ericaceae, the scientific name for the Heath family, was derived from a Greek name for a now unknown heath. The family is made up of 140 genera and approximately 3,000 species of shrubs and trees. Most prefer acidic soil and the cooler temperatures of temperate zones, growing often in moist shaded areas, along streams and on mountain slopes. Although the leaves are usually opposite, they may be alternate or whorled. The flowers have four or five sepals and four to seven petals. The petals are often joined at the base to form a tube so that the flowers are urn- or bell-shaped. Many members of the family are commercially important for their fruit (*Vaccinium* spp.: blueberries, cranberries and related shrubs; and *Gaylusacia* spp.: the huckleberries*); others are important as decorative garden plants including species of *Rhododendron* (a genus that now includes Labrador tea and azaleas). Four members of the Heath family are state flowers, including *Rhododendron maximum* for West Virginia and *Rhododendron californicum* for Washington. The mountain laurel *(Kalmia latifolia)* is Pennsylvania's state flower, and the fragrant and secretive mayflower or trailing arbutus (*Epigaea repens*) is the state flower of Massachusetts.

---

*There is considerable local variation in the use of common names for the berries in these two genera, and many names are used: blueberries, huckleberries, bilberries, whortleberries, etc. The recent tendency is to place the huckleberries in genus *Gaylussacia;* the blueberries continue as *Vaccinium.*

Rusty menziesia
Menziesia ferruginea Sm.

Rusty menziesia (also known as false azalea and fool's huckleberry) honors Archibald Menzies, who first collected the plant. It is an attractive deciduous shrub that grows from California to Alaska, and inland to the northern Rocky Mountains. The plants prefer moist surroundings, and are often found in the company of trapper's tea (*Rhododendron neoglandulosum* Harmaja). Urn-shaped flowers, ranging in color from rust to bright red, help to identify the plant. The fruit, while resembling that of various species of blueberries and huckleberries, is inedible, hence the common name "fool's huckleberry." Menziesias are sometimes used as ornamental shrubs in moist situations.

Trapper's tea
Rhododendron neoglandulosum * Harmaja

Trapper's tea (formerly *Ledum glandulosum* Nutt.) is a western subalpine to alpine woody plant that blooms in early summer. It is commonly found on the shores of high mountain lakes along slow-moving streams and on swampy ground. While the plant has always been classified as a *Ledum,* recent genetic studies show that its proper taxonomic classification is as a species of *Rhododendron* (although the name "ledum" persists and will probably be retained as a common name for some time). It is easily identified by where the plants grow as well as by its large, white flowerheads set off by deep green, leathery leaves. The fruit is an inconspicuous brownish capsule that contains numerous seeds. Its distribution is centered on Idaho, for it is found in every contiguous state and Canadian province as well as in California and the province of Alberta. The derivation of the common name "trapper's tea" is uncertain, but the leaves presumably were used for tea in the distant past. Readers should not emulate our pioneer forebears, because the plant is poisonous.

---

*Since there was already a plant named *Rhododendrum glandulosum,* it was necessary, according to the rules of botanical nomenclature, to give the erstwhile ledum its own, unique species name (its "specific epithet"), hence *Rhododendron neoglandulosum.*

Grouseberry (left)
Vaccinium scoparium
Leiberg ex Coville

The grouseberry (also known as grouse whortleberry) is a small mountain shrub that often forms large patches of ground cover at high elevations. Its range extends from Alberta and British Columbia, throughout our Northwest, south to California and New Mexico and east to Montana and South Dakota. The plants have many thin branches that when bundled form a serviceable broom (*scoparium* means "broom-like"). In good years the plants bear many little flowers typical of those of other *Vaccinium* species (blueberries, cranberries, etc.). These ripen into the red berries shown here. They are sweet and taste exactly like other species of blueberries. Other blueberries do grow in our mountains, but the grouseberry is the most common subalpine species.

Kinnikinnick (above)
Arctostaphylos uva-ursi
(L.) Spreng.

Kinnikinnick (also known as bearberry) is a circumboreal plant found throughout the northern United States and Canada, north to Alaska and south to Arizona and Virginia. It is a ground-hugging plant that forms tight mats of oval leathery leaves. Urn-shaped, early-blooming white flowers mature into bright red berries. The fruit is not really palatable, but Native Americans apparently used it as a component of pemmican. A fur trader gave the explorers Lewis and Clark a specimen of kinnikinnick while they were overwintering at the Mandan villages in today's North Dakota (1804–1805). He told them that the Indians mixed leaves of kinnikinnick with the bark of the red-osier dogwood (page 92) and tobacco to stretch the tobacco supply. Kinnikinnick grows in our mountains mostly from the Salmon River northward, in moist or shady locations, although we have also seen it growing on alpine tundra farther south.

Merten's mountain heather
Cassiope mertensiana (Bong.) G. Don
var. gracilis (Piper) C. L. Hitchc.

The three mountain plants shown here are subalpine and alpine plants that grow in high, wet meadows and on the banks of mountain lakes, often in each other's company. This mountain heather and the pink mountain heather shown below are so closely related that they often hybridize when growing close to each other. Both plants are native to western coastal and Rocky Mountains states (although this variety is found only in Oregon, Idaho and Montana). The name *Cassiope* is from Greek mythology, the name of Andromeda's mother. The species name honors German botanist Franz Karl Merten (see page 66) whom we have met earlier in this book.

Pink mountain heather
Phyllodoce empetriformis (Sm.) D. Don.

Both this and the mountain heather shown above are low evergreen plants. This one's flowers are pink rather than white and its joined petals roll outward along their margins. Phyllodoce was a water nymph in Greek mythology. Why the names of mythological Greek characters were applied to alpine heathers is a mystery; *empetriformis,* from the Greek, means "on rocks." Frederick Pursh in his 1813 *Flora* reported seeing a Lewis and Clark specimen that was probably gathered in north-central Idaho (the present location of the specimen is unknown). Attempts to cultivate this lovely little alpine plant have always ended in failure.

Alpine laurel
Kalmia microphylla (Hook.) A. Heller

Our alpine laurel is closely related to the state flower of Pennsylvania, *Kalmia latifolia*, a white-flowered shrub or small tree often used in ornamental landscaping. Our plant is only a few inches tall—an adaptation to its alpine surroundings. The relationship between the two plants is confirmed by this plant's woody stem and almost identically shaped flowers. As with the two mountain heathers shown above, the alpine laurel prefers wet mountain meadows and bogs, where it blooms soon after snowmelt. Its distribution is similar to that of the other two plants. Alpine laurel is poisonous to cattle and sheep—not usually a problem, given the altitude at which it grows.

Pink wintergreen (left)
Pyrola asarifolia Michx.

This generic name, *Pyrola,* was derived from the Latin *pyrus* for "pear," because of similar shaped leaves in some species of wintergreen. The name *asarifolia* suggests that the leaves resemble those of the wild ginger (*Asarum caudatum*). The pink wintergreen is a reclusive plant, often found at higher elevations. Nodding pink flowers form a loose cluster atop a bare stem. Each has five dainty pink petals, ten stamens and a single long, protruding style—an aid to identification.

Green wintergreen (right)
Pyrola chlorantha Sw.

As both common and scientific names suggest, the green wintergreen has light green-petaled flowers (*chlorantha,* from the Greek, means "green flower"). It grows in partial shade, often in conifer forests. In common with the other wintergreens, it blooms in midsummer.

White-vein wintergreen (left)
Pyrola picta Sm.

While this plant's flowers are similar to those of our other pyrolas, its patterned basal leaves make it stand out in the shaded forests where it grows. Most of the plants in the wintergreen family grow all across North America, but this plant is found only in the West and in the Black Hills of South Dakota.

Sidebells wintergreen (right)
Orthilia secunda (L.) House

This little plant, while similar to the pyrolas, has somewhat different flowers. It is easily identified because its greenish-white flowers grow on only one side of the stem (*secunda,* from the Latin, means "turned," or, in this context, "one-sided"). It is a common plant and will be found blooming in mid- to late summer. As the name "wintergreen" implies, all these plants are evergreen.

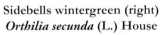

Pipsissewa
Chimaphila umbellata (L.) W. P. C. Barton

The pipsissewa, or prince's pine, is a shade-loving, evergreen plant that bears attractive parasol-shaped, five-petaled pink flowers. As with our other wintergreens, it grows in cool forest surroundings. The plant is found throughout the West, in the northeastern states, in the Canadian provinces and in Eurasia. The genus name, *Chimaphila*, means "winterlover," derived from its evergreen properties. This and other wintergreens have long been used medicinally both topically and in beverages, probably without any real therapeutic benefit.

A related plant, the little prince's pine, *Chimaphila menziesii* 'R. Br. ex D. Don Spreng. (not shown), occurs in the mountains of central and northern Idaho. It has few flowers, and its leaves tend to be broadest near the base of the blade.

One-flower wintergreen
Moneses uniflora (L.) A. Gray

The one-flower wintergreen (also known as single-delight, wood nymph, shy-maiden and wax-flower) is a widely distributed plant that grows in all our northern states, in the Rocky Mountain states, throughout Canada and in Eurasia. It grows in moist places as high as the subalpine zone. With its single (*uniflora*), five-petaled, nodding white flower and rosette of basal leaves, it is easy to identify. It is the only species in its genus. The name *Moneses* was derived from two Greek words, *monos* and *hesis*, meaning "single delight."

Indian pipe
Monotropa uniflora L.

The Indian pipe occurs in Idaho, although it is seen less commonly than the pinesap shown below. It is native to our northwestern states, Alaska, across Canada and in all the eastern states. The two plants shown on this page are both saprophytes that thrive on forest litter. Plants in this genus lack chlorophyll, and their root systems are always associated with minute fungi. The generic name, *Monotropa*, was derived from the Greek words *mono* for "one" and *tropos* for "direction," evidently referring to plants that bear their flowers on one side of the stem.

Pinesap
Monotropa hypopithys L.

The pinesap is also found in most of the United States and in Eurasia. It is easily distinguished from the Indian pipe by its yellow color, which turns brown as it matures. Several small flowers are clustered at the top of the stem, compared to the Indian pipe's single flower. The species name, *hypopithys*, was derived from two Greek words, *hypo* for "under" and *pithys* for "pine," alluding to the pine forests in which the plants are usually found. Identification is usually not a problem, although this plant and species of *Corallorhiza* in the orchid family are rather similar in appearance, and both are saprophytes that grow on shaded leaf-litter.

Fabaceae

Pea Family (Fabaceae)

The older scientific name for the Pea family, Leguminosae, is one of several family names that don't end in "-aceae," so many botanists prefer to use Fabaceae (from the Latin word *faba* for "bean"). Either name is correct. The family is a large one and taxonomically difficult even for botanists. It is made up of more than 630 genera and approximately 18,000 species including herbs, vines, shrubs and trees. The Pea family is second only to the grasses (Poaceae) in economic importance. The flowers of many of the family's plants are made up of five petals: these include a large upper petal, (banner), two smaller lateral ones (wings), and the two lowest ones joined together to form a keel. The flowers are commonly referred to as being "papilionaceous," a word derived from *papilion*, the Latin word for "butterfly." The fruit is a pod that splits open along two seams. Leaves are mostly compound: either pinnate (like a feather) or palmate (leaflets all arise from a central point like fingers from the palm). The family includes beans, peas, lentils, peanuts, clover, alfalfa, etc., plants important not only for their food value to man and domestic animals, but also for their ability to fix soil nitrogen. About twenty-five genera, including both native and introduced species, grow in the Northwest—many in our mountains. Although some have edible fruit, others are poisonous; it is best to regard all wild leguminaceous plants as inedible.

Bessey's crazyweed
Oxytropis besseyi (Rydb.) Blank
var. *salmonensis* Barneby

Bessey's crazyweed is one of the more colorful members of the pea family. It is characterized by gray-green, furry leaves and long stems topped with tight clusters of brightly colored red to reddish-purple flowers. All parts of the plant save the flower petals are noticeably hairy. Half a dozen varieties of Bessey's crazyweed are recognized, varying by differences in size, distribution, form of the hairs, etc. While the species as a whole is relatively common, the variety shown here is not. It is found only in Idaho where it grows in dry mountain valleys (e.g., Malm Gulch) that drain into the Salmon River south of Challis. The species *Oxytropis besseyi*, in all its half dozen varieties, ranges from Alberta and Saskatchewan (where it is rare), south through Idaho, Montana, Wyoming, Nevada, Utah and Colorado (where it is also uncommon).

Sharon Huff

Pursh's milk-vetch
Astragalus purshii Douglas ex Hook.
var. *concinnus* Barneby

This milk-vetch, *Astragalus purshii*, grows as high as the montane zone. There are many varietal forms. All have a prominent calyx, and their flowers range in color from white through yellow to a purple-tinged pink, as in this illustration. The species name honors Frederick Pursh (1774–1820), the botanist who identified and published descriptions of the plants that Lewis and Clark collected. David Douglas first gathered this species and suggested that it be named to honor Pursh. The variety shown here is native to central Idaho and western Montana; the name *concinnus* means "neat" or "elegant."

Canadian milk-vetch
Astragalus canadensis L.
var. *mortonii* (Nutt.) S. Watson

The Canadian milk-vetch is a common plant that grows from sea level to at least as high as the montane zone in our mountains. The species is found throughout North America, although the varietal form shown here grows only in the northwestern states and British Columbia. The plants spread by rhizomes to form distinctly circumscribed patches. They may also be identified by their unusually tall (for a milk-vetch) upright stems, congested clusters of yellowish-white pea-like flowers and, in our variety, by a calyx covered with black and white hairs. The variety, *mortonii*, was gathered by Nathaniel Wyeth in 1833 and named to honor Philadelphia naturalist Samuel George Morton (1799–1851).

Indian milk-vetch
Astragalus australis (L.) Lam.
var. *glabriusculus* (Hook.) Isley

Indian milk-vetch (formerly *Astragalus aboriginorum*) grows to treeline and above, on exposed, windswept, rocky soil. It is found throughout the northern Rocky Mountains and neighboring states, in the two western Canadian provinces and Alaska, as well as in Europe and in eastern Russia. The plants vary from place to place, but typically the banner is erect, the flowers are whitish and often have purple markings on the keel, and the ends of the wings are slightly notched. The leaves have seven to fifteen closely ranked leaflets. Its roots are said to be edible.

Field, or purple, milk-vetch
Astragalus agrestis S. Watson

The field milk-vetch is a common meadow plant that may grow as high as treeline and is found as far east as the Mississippi River in the United States and Manitoba in Canada. Clustered purple flowers are borne at the end of a separate stem. Its leaves are made up of twelve to twenty-three long, oval leaflets. The pod is rather short and said to be three-cornered. The species name, *agrestis*, means "rustic."

Alpine milk-vetch
Douglas ex G. Don

Despite its name, the alpine milk-vetch also grows at lower altitudes. Its distribution in the United States is much the same as the rather similar purple milk-vetch above, although it is a more widely distributed circumboreal plant that grows in all the Canadian provinces, Greenland and Eurasia. Both plants are commonly encountered in Idaho. The alpine milk-vetch's leaflets are smaller and rounder, and its pod is longer and more darkly pigmented, than those of the purple milk-vetch.

Spiny milk-vetch
Astragalus kentrophyta

The spiny milk-vetch is found as one variety or another in mountainous areas of the Dakotas, in Nebraska and all the states to the west and in the Canadian provinces of Alberta and Saskatchewan. As shown in the photograph, it forms spiny ground-hugging clumps or mats. The plant is quite variable, and nine varieties have been described.

Late yellow crazyweed
Oxytropis monticola A. Gray

The crazyweeds (also known as loco-weeds) take their common name from the poisonous effect they have on grazing animals. The scientific name, *Oxytropis*, means "sharp keel" for a pointed leading edge where two lower petals are fused. Typically, short-stemmed, papilionaceous flowers are borne on an erect stem. The leaves are made up of varying numbers of leaflets with a single terminal leaflet; they are odd-pinnate. A subalpine and alpine plant, it occurs in Washington, eastern Oregon, central Idaho, and east to Alberta, the Dakotas and Colorado. The species name, *monticola*, means "mountain-loving." The plant, also known as the mountain crazyweed, was formerly classified as *Oxytropis campestris* var. *gracilis*.

Northern yellow crazyweed
Oxytropis campestris (L.) DC.
var. *cusickii* (Greenm.) Barneby

The northern yellow crazyweed is a variable, far-flung species. Twelve varieties are recognized; var. *cusickii* is the only one found in Idaho. The plant grows as high as the alpine zone, where this plant was photographed. It is a low plant at this altitude, with odd-pinnate leaves, each with seventeen (or fewer) leaflets. The species, in its many varieties, grows all across Canada and the western United States, south to Utah and Colorado, west to Oregon and Washington and east to Minnesota. It is also native to Eurasia.

Nevada pea
Lathyrus lanszwertii Kellogg
var. *aridus* (Piper) Jepson

Lathyrus was a name used by ancient Greeks for the European chickpea. Plants in this genus are commonly known as vetchlings or sweet peas. They are rambling, climbing plants whose compound leaves typically have one or more tendrils at the end of the feather-like compound leaves (a few species lack tendrils). The Nevada pea (also known as Lanszwert's vetchling or thick-leaved pea) is often seen growing from high foothills to montane ponderosa pine forests, distinguished by large, usually white or sometimes pink-tinged flowers borne in small clusters. The plants are found in most western states. The species was named for a Belgian-born California pharmacist, Louis Lanszweert (1825–1888).

Mountain thermopsis
Thermopsis rhombifolia Nutt.
var. *montana* (Nutt. ex Torr. & A. Gray) Isley

The mountain thermopsis (or mountain golden pea) is a tall, showy plant that would be quite at home in an ornamental garden (the plants, in fact, do well when grown from seeds). Bright yellow, loosely clustered flowers and clover-like leaves help to identify it. The name *Thermopsis* comes from two Greek words: the first was used for lupines, and *-opsis* is an ending that means "looks like." Two varieties of this plant grow in Idaho. This one, var. *montana*—photographed in the Clearwater Mountains—occurs from eastern Idaho and adjacent Montana to southeastern Oregon and south in the mountain states to New Mexico. A second, broadleaf variety, var. *ovata*, occurs farther to the west and north.

Silky lupine
Lupinus sericeus Pursh

The silky lupine is usually blue, but ranges in color from off-white to the intense blue shown here. It is mainly a foothills plant, although it may grow as high as the subalpine zone. It can be distinguished from *Lupinus argenteus* (following page) by its rounded leaflets and a banner (the large upper petal) that is usually hairy on the back with a white center. The silky lupine was unknown to science until Meriwether Lewis collected a specimen on the Clearwater River in today's north-central Idaho on June 5, 1806.

It can be difficult to distinguish between *Lupinus sericeus* and *Lupinus argenteus*. The flower clusters of the former tend to be looser and are lower, extending well into the plant's leaves, with the individual flowers smaller. The banners are often lighter or partially white. Both plants are variably hairy; *Lupinus sericeus* tends to be more so, and its calyx has a hump on the dorsal surface.

Silver lupine
Lupinus argenteus Pursh
var. *depressus**
(Rydb.) C. L. Hitchc.

Although the silver (or silvery) lupine is found in most states west of the Mississippi River, the variety shown here is restricted to the mountains of Idaho, Montana and Wyoming. It seldom grows more than a foot high and is quite at home at high elevations; in August, alpine meadows are often covered with its bright flowers. It may be identified by this growth preference and by its crowded clusters of purple to blue flowers. *Lupinus argenteus* was unknown to science until the Lewis and Clark Expedition returned a specimen (var. *argenteus*), collected in Montana on July 7, 1806.

---

*Recently, some have classified this plant as its own species, *Lupinus depressus* (the species name means "depressed" or "low").

Longspur lupine
Lupinus arbustus Douglas ex Lindl.
var. *calcaratus* (Kellogg) S. L. Welsh

This plant, formerly classified as *Lupinus calcaratus* has recently been reclassified; *arbustus* means "small tree" or "shrub" and *calcaratus* means "spurred" for a bump-like projection that extends backward from the top of the calyx. It is common in the Rocky Mountains and west to the coast, often growing in great profusion on sagebrush slopes as high as the subalpine zone. The color of the flowers varies considerably, from light purple to yellow on the same plant—those shown here are typical.

Stemless dwarf lupine
Lupinus lepidus Douglas ex Lindl.
var. *utahensis* (S. Watson) C. L. Hitchc.

This little lupine grows in the grass of montane meadows. Its flowers have short stems, and the leaves, stems and base of the flowers are covered with long hairs, giving the plant a furry, grayish appearance that helps to identify it. The "stemless" in the common name refers to the leaves and flowers, not to the plant itself.

Lupinus lepidus grows in Idaho, Oregon, Washington and, uncommonly, farther north in British Columbia and Alaska. The name *lepidus*, from the Latin, means "neat" or "charming." A standardized common name, "elegant lupine," has been suggested for the plant.

Owyhee clover
Trifolium owyheense Gilkey

The attractive little owyhee clover is a rare perennial plant that grows only in a localized area of southwestern Idaho and adjacent Oregon. While the chances of encountering the plant are slim, it is worth looking for it in late spring. (It grows south of Marsing off US Highway 95. Look in the vicinity of Succor Creek or Leslie Gulch.)

Long-stalked clover
Trifolium longipes Nutt.

Several native clovers grow in our mountains. The long-stalked clover is the species most frequently seen at higher elevations, where it grows as high as the alpine zone. It is easily identified by its single small flowerhead atop a rather tall stem (*longipes* means "long-stalked"). Close to a dozen varieties are recognized, and their classification is based on minor technical differences. Its trifoliate leaves bear lance-shaped leaflets.

Purple-flowered woolly vetch
Vicia villosa Roth

This purple-flowered plant, also known as winter vetch, is a colorful import that is now fully naturalized. It is often seen growing near roadsides at higher elevations. It is villous (hairy), and its flowers grow in a spike-like cluster, always on one side of a tallish stem—an identifying feature. Another species of *Vicia*, the common purple-flowered vetch or tare, *Vicia sativa* L. (not shown), is often seen near cultivated fields. It is characterized by large flowers and dainty vine-like foliage whose pinnate leaves end in tendrils that clasp any nearby vegetation, usually sagebrush in our area.

Fabaceae

White clover
Trifolium repens L.

The white clover is an everywhere plant, introduced in the long-ago past; it is now found in every state and province in North America. Its species name, "*repens*," is often used in scientific names for creeping plants (cf. "reptile"), describing a tendency for the plant to spread along the ground. The trifoliate leaves are small and often form patches of considerable size in moist situations. It is seen, not uncommonly, growing in high mountain meadows far from settled areas.

Yellow sweet-clover
Melilotus officinalis (L.) Lam.

Yellow sweet-clover (above center) and the very similar white sweet-clover (above right) were formerly classified as separate species (the white clover as *Melilotus alba* L), but now both are considered to be varieties of the same species, *Melilotus officinalis*. They were introduced long ago from Europe to the New World as fodder plants. They are now found throughout North America and are listed as weeds in many places. They are not without value, however, for they fix soil nitrogen. They spread rapidly and may grow to be bushy. The flowers and leaves are tiny. Sweet-clovers are extremely common along our roadsides, at all elevations. The name *Melilotus* is an ancient term meaning "honey-clover"; Homer is said to have used it to describe animal feed.

Pea Family Imports
A few nonnative members of the Pea family are shown on this and the following page. These (and many others) are now fully naturalized and are often seen, usually near agriculturally developed areas.

Red clover
Trifolium pratense L.

The red clover is an introduced species, useful both as a cultivated forage plant and as an ornamental. A Eurasian import, it is now well established throughout North America. The red clover is easily identified by its large trifoliate leaves, ovoid leaflets and a large—sometimes very large—reddish-purple flowerhead. The species name, *pratense*, means "of the meadows." It is at home in our mountains, growing in open places as high as the subalpine zone.

Sainfoin
Onobrychis viciifolia Scop.

Sainfoin is a European import whose usefulness as fodder has been known for millennia. In fact, the name *Onobrychis,* from the Greek, means "donkey food." "Sainfoin," in turn, is derived from the French words *sain[t]* and *foin*, the latter word means "hay," thus "blessed hay." Its flowers are borne on a tall stem; its feathery leaves (*viciifolia* means "vetch-like leaves") have small, opposing leaflets. The flowers are small, but showy, with petals that are more prominently veined than those of other members of the Pea family that grow in our area. Sainfoin, in profusion, sometimes turns montane hillsides pink.

Alfalfa
Medicago sativa L.

There is no problem identifying this common plant; for anyone who grew up in a rural environment, it is obviously alfalfa. Others, however, may not recognize it on first encounter. Alfalfa commonly escapes and may show up at a considerable distance from the field in which it was planted (the Latin word *sativa* means "sown" or "cultivated"). Unlike many introduced members of the Pea family, alfalfa is a valuable forage plant, used either as harvested or dried as hay. Its usefulness as feed is not confined to farm animals; there are some who profess a fondness for alfalfa sprouts in salads and sandwiches.

Fumitory, or Bleeding Heart Family (Fumariaceae)

The Fumitory family is not a large one, consisting of only sixteen genera and 515 species. The name Fumariaceae is derived from that of a common European plant, *Fumaria officinalis*, found in the United States only as an ornamental plant or garden escape. Members of the genus are known as fumitories, a term derived from the Latin terms *fumus* for "smoke" and *terrae* for "of the earth," possibly because the European plant has a wispy, amorphous growth pattern. Because the fumitory is little known here, the family is often referred to as the Bleeding Heart family for a plant well known to Americans. Flowers in this family are bilaterally symmetrical and four-petaled. Because the flowers often take unusual forms, the petals may be hard to delineate—as in the plants shown on these pages.

Cusick's fitweed
Corydalis caseana A. Gray var. *cusickii* (S. Watson) C. L. Hitchc.

Cusick's fitweed (or corydalis) is an uncommon, high montane to subalpine streamside plant that blooms in early summer. Its small flowers (inset, right) are unusual. An upper petal forms a long backward spur that extends as a hood over the front of the flower. This, joined to a lower petal, creates a flower tube that contains two darker inner petals. A tight cluster of fifty or so of these small purple flowers tops each of the plant's long stems. All in all, this is an impressive plant. The genus is variable and several varieties are recognized; some are quite rare. In one or another of its varieties, Case's fitweed is found in Idaho, Washington, Utah, New Mexico and, rarely, in Washington, Colorado and California. Sheep that graze on fitweeds die in convulsions, explaining its common name. Botanist Asa Gray named the species in 1874, giving it the species name *caseana* to honor Eliphalet Lewis Case (1843–1925), a California schoolteacher, Civil War veteran and occasional plant collector who found the plant.

Golden corydalis
Corydalis aurea Willd.

The golden corydalis is an attractive sprawling plant with short-lived yellow flowers. It is occasionally encountered in our mountains, growing at least as high as the montane zone. Like other members of the genus, the golden corydalis can be identified by its tubular flowers. Each contains two inner petals. One of the outer petals forms a fringed crest over the opening of the flower tube and extends backward to form a prominent spur. The plant's grayish-green leaves are basically pinnate, although the leaflets divide again (i.e., they are bipinnate). Corydalises have exploding seed capsules that effectively broadcast many small seeds for a long distance.

The golden corydalis grows in most of our western and northern tier states and throughout Canada. Unusually, for such a wide-ranging plant, no subspecies or varieties are recognized. An inelegant standardized name, "scrambled eggs," has been proposed for the plant. *Corydalis* is the Greek name for the European crested lark; presumably the name is attached to this flower because of its spur (analagous, perhaps, to "larkspur," the common name for delphiniums), or possibly for the flower's prominent crest.

Steer's head
Dicentra uniflora Kellogg

This little plant, with horns formed by two upper petals and an elongated head formed by two lower ones, is aptly named. It blooms early in the spring at montane to subalpine elevations. Steer's heads are not rare, but they are elusive. Look for them on rocky hillsides below snowline. They are small and often grow in sagebrush, so finding one is usually a matter of chance. A similar plant (*Dicentra pauciflora*) with shorter horns grows in California, so ours is sometimes known as the "longhorned steer's head."

Several other unusual wildflowers (not shown here) also belong to *Dicentra*. These include bleeding heart (*Dicentra formosa* [Haw.] Walp.) and dutchman's breeches (*Dicentra cucullaria* [L.] Bern.). Both are native to the Northwest and both are used as ornamental plants.

Gentian Family (Gentianaceae)

Gentians are the aristocrats of alpine flowers. Finding a patch of bright blue gentians blooming at summer's end is ample reward for climbing high. Many gentians favor moist, rich soil deposited on the banks of slow-flowing streams, along the shores of mountain lakes, or in wet mountain meadows. Late-blooming gentians are tolerant of cold and may continue to flower even when the nights are freezing, for they are low to the ground and sheltered by sun-warmed earth—microclimate is everything for mountain flowers. Gentians have been so-named for more than two millennia. King Gentius in the second century BC ruled Illyria, a Balkan country. The king believed that the bitter roots of a certain flowering plant had medicinal value. He was wrong, but the plant has been associated with his name ever since, and so we have "gentian" today. The family is made up of seventy-nine genera and about 1,270 species. Most are found in the north temperate zone; a half dozen or so are cultivated as ornamental plants. Species belonging to at least five genera, *Gentianopsis, Gentiana, Gentianella, Swertia* and *Frasera,* grow in the mountains of Idaho. Typically their flowers are symmetrical and their leaves are opposed. The petals are joined at the base to form trumpet, urn-shaped or salverform blooms; *Gentianopsis* and *Frasera* flowers have four petals, and other genera have five.

Explorer's gentian
Gentiana calycosa Griseb.

The explorer's gentian (also known as the mountain, or bog, gentian) is the largest, showiest, highest and latest blooming of our gentians. It grows in moist subalpine meadows and along the loamy banks of slow-flowing streams. Its flowers are five-petaled, speckled on the inside surface and have a "pleat" connecting the petals. The roundish leaves are wider than those of our other *Gentiana.* Its species name, *calycosa,* from the Latin, refers to the underlying calyx that cups the flower parts. Explorer's gentians grow in every state contiguous to Idaho, in California and in the Canadian provinces of Alberta and British Columbia, where it is classified as a rare plant.

| **One-flowered fringed gentian** | **Pleated gentian** | **Autumn dwarf-gentian** |
| *Gentianopsis simplex* (A. Gray) Iltis | *Gentiana affinis* Griseb. | *Gentianella amarella* (L.) Boerner |

One-flowered fringed gentian
***Gentianopsis simplex* (A. Gray) Iltis**

The one-flowered gentian has a small and—as its species name suggests—a plain flower, noticeable mostly for its intense blue color. Until recently classified in genus *Gentiana*, it is now in *Gentianopsis*, a genus made up of similar four-petaled gentians. It is a close relative of the Rocky Mountain fringed gentian *Gentianopsis thermalis* (Kuntze) Iltis. (not shown), a plant also found in Idaho. Both have four petals, but this one usually has little if any fringing at the end of the petals. Although not shown in the image, the leaves are opposed and narrow. Both plants grow along streams and in moist, high montane to subalpine meadows, blooming from mid-July through August. The species grows in Idaho, west to Oregon and California and, rarely, in Montana and Wyoming.

Pleated gentian
***Gentiana affinis* Griseb.**

The pleated gentian takes its common name from an inward-folding membrane that joins one petal to the next. As in the plant shown, the petals often have a reticulated pattern of greenish spots on their upper surfaces. Pleated gentians spread by sending out roots, and are often found growing in clusters that sometimes suggest the "fairy rings" formed by proliferating field mushrooms. Each of the pleated gentian's stems bears several pairs of small, opposite, lanceolate leaves and a terminal flower. As with other members of this genus, this gentian prefers moist montane meadows and the banks of lakes and streams. The pleated gentian is a well-distributed species, found throughout the western United States and across Canada to Ontario.

Autumn dwarf-gentian
***Gentianella amarella* (L.) Boerner**

The autumn dwarf-gentian is the daintiest of our gentians. It flowers during the second half of August in moist, high montane to subalpine meadows. It is a circumboreal plant that, with minor differences, ranges across North America, where it is found in all our western states and Canadian provinces, to Europe and Asia. Our plant is pink, but colors vary from white to pale blue or purple. The plant may be identified by a fringe of prominent hairs inside the flower—shown clearly in the magnified view above. The species name, *amarella*, from the Latin, means "a little bitter."

115

Clustered elkweed (above)
Frasera fastigiata (Pursh) A. Heller

The clustered elkweed grows only at the base of Idaho's panhandle from Idaho County north to Latah County and west to the nearby Blue Mountains in Oregon and Washington. It is a tall, large-leaved plant topped with four-petaled blue flowers. It is the only member of the gentian family that Meriwether Lewis collected on, June 14, 1806, on the Weippe Prairie. The species name, *fastigiata*, implies "cone-like" for the shape of the flower cluster.

White gentian (above)
Frasera montana Mulford

The white gentian is localized to central Idaho, found north of Galena Summit (Idaho Highway 75). It is common around Stanley and Redfish Lake and is found as far to the west as Boise County. It is a high montane plant that blooms in late spring on dry ground, usually in the company of sagebrush. Four-petaled, clustered flowers and its white-bordered, narrow leaves, especially, help to identify it.

Monument plant (center)
Frasera speciosa Douglas ex Griseb.

The monument plant (also known as giant frasera or green gentian) is a tall, narrow, cone-shaped plant with flowers clustered around the upper part of its stem. The plants live for many years, but bloom only once and then die. Flowering is unpredictable, but seems related to moisture. The flowers are about three-quarters of an inch in diameter with long sepals between each of its four purple-spotted petals. Each petal has two pits at its base and four stamens surrounding a one-seed ovary. The name *Frasera* honors John Fraser (1750–1811), a Scots nurseryman who collected in southeastern North America. The name *speciosa* means "showy." The plant grows throughout the western United States.

Swertia
Swertia perennis **L.**

The swertia (also known as felwort or star gentian) is the only member of its genus found in America. It is a late-blooming (i.e., end of August) circumboreal plant found in most of the western states, in British Columbia and north to Alaska (although rarely encountered there). The plants also grow in Eurasia, as the "L." after the binomial scientific name suggests, for it stands for Linnaeus, who may have found it during his exploration of Lapland. Swertias favor wet montane to subalpine meadows. The flowers' four petals range in color from a light to a dark purple, the latter so deep as to be almost black. Longitudinal veining of the petals, not especially prominent here, may be quite pronounced. The flowers, magnified in this photograph, are about three-quarters of an inch across. Each has a prominent fruiting capsule standing erect in the center. The petals have two nectar pits at the base, although it takes magnification to see these. Emanuel Sweert (1552–1612), whom Linnaeus honored with the plant's name, was a Dutch botanist who composed a catalog of plants.

Geranium Family (Geraniaceae)

The Geranium family consists of five genera and 760 species. The name is derived from the Greek word *geranos*, meaning "crane," referring to the long pointed beak formed by the style as the seeds form. Perversely, the popular and attractive garden "geranium" does not belong to the genus *Geranium* at all, but to *Pelargonium*, a genus made up of almost 300 tropical plants found mostly in South Africa. The eco-nomic importance of the family is due entirely to the popularity of its cultivated ornamental varieties. Those in genus *Geranium* differ from the pelargoniums by the symmetrical configuration of their flowers, with five sepals, five petals, ten stamens and a five-part pistil, and their (usually) pink to purple coloration. About ten species of *Geranium* grow in the Northwest, but of these, only two are native species.

Sticky geranium, *Geranium viscosissimum* Fisch. & C. A. Mey.

The sticky geranium flowers from spring until nearly summer's end. The flowers are usually pink with deep red to purple veins, although their color varies from almost white to light purple. Deeply cut compound leaves are a distinguishing feature. The plants prefer, but are not restricted to, moist and shady areas; they grow as high as the subalpine zone where the plant's morphology changes as they become prostrate. The leaves and stems are sticky, explaining both common and scientific names. The very similar white geranium, *Geranium richardsonii* Fisch. & Trautv. (not shown), also a native species, grows in a like environment farther north in Idaho than this plant. Both are found throughout the West.

Crane's bill, *Erodium cicutarium* L'Hér. ex Aiton

The crane's bill (also known as stork's bill and filaree) is a small, pink-flowered Eurasian plant that now has spread worldwide and is found almost everywhere on the North American continent. In early spring the plants often appear in great numbers. Although considered a weed, the erodium is an excellent browse plant. It grows as high as the montane zone, although is less common there than at lower elevations. Crane's bills are creeping plants, usually only an inch or so high. Their long, persistent styles explain the erodium's scientific and common names (*Erodium,* from the Greek, means "crane" or "heron"). The species name, *cicutarium,* implies that its leaves resemble those of the *Cicuta,* the European water hemlock.

Currant Family (Grossulariaceae)

The Currant family is a small one, consisting of only one genus and about 200 species. In the past, currants and gooseberries were included in the Saxifrage family, but are now in their own family, Grossulariaceae. The currants (*Ribes*) and the gooseberries (*Grossularia*) were also classified as two separate genera, but the plants are so similar that both are now included in a single genus, *Ribes* (pronounced RIBE-eez, from an Arabic word that means "acidic"). The main difference between currants and gooseberries is that the latter have prickly (armed) stems, whereas currants are smooth-stemmed. All members of the currant family are shrubs that bear (mostly) edible berries. The garden gooseberry (*Ribes uva-crispa*) and the cultivated currant (*Ribes nigra*) have been cultivated for centuries. Many wild species of *Ribes* grow in our mountains, and they resemble each other—and the domesticated fruit—to the point that if one is acquainted with one species it is not difficult to recognize others as members of the same family.

Henderson's gooseberry
Ribes oxyacanthoides L. var. *hendersonii*
(C. L. Hitchc.) P. K. Holmgren

Henderson's gooseberry is an alpine plant that is restricted to Idaho's Lost River Range (where this plant was photographed), and Montana's Anaconda and Nevada's Toiyabe mountain ranges. Although it lacks small branch prickles, it is armed with impressively long thorns. We have not seen a fruiting plant. Several other varieties of *Ribes oxyacanthoides* are recognized; these occur at lower elevations. Louis Fourniquet Henderson (1853–1942), for whom this variety was named, was professor of botany at the University of Idaho and, subsequently, at the University of Oregon, during the first half of the twentieth century.

Golden currant
Ribes aureum Pursh

The golden currant grows in all but our southeastern states. It is identified by its yellow flowers, three-lobed leaves and tasty berries. While all *Ribes* are to a degree edible, the golden currant is the sweetest. The Lewis and Clark Expedition first collected the plant near the Three Forks of the Missouri on July 29, 1805, and again the following spring near today's The Dalles on the Columbia River on April 16, 1805.

Squawberry
Ribes cereum Douglas

The squawberry, or wax currant, grows as high as treeline in our western states, and east to South Dakota and Oklahoma. Its creamy-white flowers are about half an inch long, and its translucent berries are edible, but tasteless. Indians used the fruit for pemmican, hence the plant's common name. David Douglas was the first to collect this species along the Columbia River in 1825. The name *cereum* means "waxy" for the berries' appearance.

Swamp black gooseberry
Ribes lacustre (Pers.) Poir.

This prickly wild gooseberry is another easily identified *Ribes*. It grows in moist places and on streambanks. The small, filmy flowers and shiny leaves are distinctive. Its stems are prickly. The plant's black fruit may also have soft prickles, unlike those of other wild currants. The berries are edible but sour—only suitable as an emergency food.

Sticky currant
***Ribes viscosissimum* Pursh**

Touch this plant's foliage and you will see how it got its common and scientific names. The berries are black although we have yet to encounter a fruiting plant. Meriwether Lewis collected the plant—unknown then to science—on June 16, 1806, on the Lolo Trail while eastward bound, noting that it grew on "The hights of the rocky mountain… Fruit indifferent and gummy…" The plant is common in our western states and adjacent Canadian provinces.

Hudson's Bay currant, *Ribes hudsonianum* Richards.

The Hudson's Bay currant (also known as northern black currant) is a montane to subalpine streamside plant that grows all across northern North America. The shrubs bear sprays of white flowers that ripen into black fruit. Maple-like leaves give off an acrid odor similar to that of cat urine, explaining another common name, "stinking currant." Despite the odor, the berries are reasonably palatable. Two varieties occur in Idaho; ours is var. *petiolare* (Dougl.) Jancz., a southern variety. The other, var. *hudsonianum,* grows near the Canadian border and farther north where it is found in all the Canadian provinces. The Hudson's Bay currant prefers streambanks, which is where they are usually found.

Hydrangea Family (Hydrangeaceae)

The Hydrangeaceae, a small family, consists of sixteen genera and about 250 species. Most grow in the north temperate zone. The family's chief economic importance lies in plants that are cultivated as ornamental trees and shrubs—the best known are species of *Hydrangea* and the mock oranges (*Philadelphus* spp.). Only one member of the family, *Philadelphus lewisii,* the state flower, is native to Idaho. While Idahoans universally refer to the plant as the "syringa," it correctly should be known as Lewis's mock-orange. It is an attractive, fragrant, white-flowered shrub or small tree that was unknown to science until Meriwether Lewis collected two specimens on the expedition's return trip in 1806. The plant came to be known as a syringa, seemingly because it was confused with the common lilac, *Syringa vulgaris,* a member of the Olive family (Oleaceae). Recent DNA studies suggest that the Hydrangeaceae family is closely related to the Cornaceae, the Dogwood family.

Lewis's mock-orange
Philadelphus lewisii Pursh

Meriwether Lewis collected the mock-orange on today's Clearwater River on May 6, 1806, and again on July 4, 1806, on today's Bitterroot River. He noted, correctly, that the plant might be a *Philadelphus.* The plant, known in Idaho as the "syringa" is not hard to identify, for its fragrant flowers set it apart from the western dogwood, the only other tree with four-petaled flowers with which it might be confused, (the dogwood's "petals" are actually bracts, whereas this plant's "petals" are petals). Apparently the mock-orange was more common when Lewis found the plant than it is today. It still grows along the Clearwater, but in reduced numbers. It is not difficult to find one today, however, for they are commonly planted as landscape ornamentals.

The generic name, *Philadelphus,* was derived from two Greek words, *philos* for "love" and *adelphos* meaning "brother." It is said to have been derived from the name of Ptolemy Philadelphus, an Egyptian king (283–247 BC).

Waterleaf Family (Hydrophyllaceae)

The Latin *hydrophyllum,* from which this family's name was derived means "water-leaf." The original waterleaf was a European plant represented in antiquity by an "ornament used in sculptured capitals, supposed to be a conventionalized representation of the leaf of some aquatic plant" (OED). Linnaeus appropriated the name *Hydrophyllum* and applied it to our plant. Hydrophyllaceae are found only in the New World. It is a relatively small family made up of fifteen genera and about 300 species. Its members are in many ways similar to those in the Borage (Forget-me-not) family—so much so that some taxonomists lump the two families together as Boraginaceae. Hydrophyllaceae are found throughout the American West and are strongly represented in the Northwest. Several members have been cultivated as ornamental garden varieties (primarily species of *Phacelia*), the family's only economic importance. Typically, flowers of the waterleaf family have five sepals and five petals that are united at their base to form small bell- or saucer-shaped flowers. These may be solitary or form clusters, many of which arise from fiddle-head-shaped helicoid cymes.

Purple-fringe *Phacelia sericea* (Graham) A. Gray

The purple-fringe is one of our more spectacular alpine flowers. It blooms in early to midsummer on the banks of high mountain lakes and in moist open areas on ridges where nearby snowbanks and cornices are still melting. The species name, *sericea*, means "silky" for the hairs on the foliage. Two varieties are recognized; ours is var. *sericea*. A larger variety, *Phacelia sericea* var. *ciliosa* Rydb., also grows in Idaho at lower elevations although we have not encountered the plant. The purple-fringe is distributed throughout the West and north to Alaska. *Phacelia* is a large genus made up of about 150 species. It is well represented in Idaho and surrounding states. Many, like the purple-fringe, are showy plants.

Silverleaf phacelia
***Phacelia hastata* Douglas ex Lehm.**
Phacelia, from the Greek *phakelos,* means "bundle" for the clustered flowers common to this genus; *hastata* means "spear-shaped" for the plant's (inconstant) pointed lower leaves with small lateral lobes. Tightly coiled helicoid cymes are prominent. At lower altitudes, silverleaf phacelias are tall, topped with unrolling clusters of small purple flowers (bottom, right). At higher altitudes the plants are prostrate and hug the ground (top, photographed at treeline), although the flowers themselves retain the same appearance (bottom, left). In the past the change in the plant's morphology with altitude caused it to be classified erroneously as a separate variety, var. *alpina* (Rydb.) Cronquist. Silverleaf phacelias grow in most of the western states and the two western Canadian provinces (although it is classified as a rare plant in Alberta).

Variable-leaf scorpion weed
Phacelia heterophylla Pursh

Phacelia heterophylla is a common plant that appears in late spring to early summer. It is impressive to look at, for it stands two feet or more high and is covered with whorls of unrolling stemlets bearing attractive white flowers. It has been classified as a variety of silverleaf phacelia in the past, but there is enough difference between the two plants to warrant placing them in separate genera. The plant's distribution neatly fills a map of the West as it ranges south from Montana to Arizona and all the states in between to the coast. While the plant has little importance as an ornamental, it is considered a major range plant important as forage for both wild and domestic animals.

Franklin's phacelia
Phacelia franklinii (R. Br.) A. Gray

The five-petaled, light purple flowers of Franklin's phacelia unroll as a scorpioid cyme. Its leaves—not shown clearly here—are pinnate with blunt lobes. The plants may have only a single stem, but more commonly there is a central stem with several smaller ones surrounding it to form a clump. The species name, *franklinii*, honors Sir John Franklin (1786–1847), an ill-fated explorer of northern North America and the arctic who died while searching for a Northwest Passage. Franklin's phacelia ranges northward from the mountains of Utah and Wyoming, through the American Northwest to Alaska and as far east as Manitoba.

Idaho phacelia
Phacelia idahoensis Henderson

The Idaho phacelia is a tall plant whose purple flowers are borne on a spike-like stem. As with other phacelias, the flowers are on coiled stemlets that gradually unroll, explaining the common name "scorpionweed" that is sometimes used for plants in this genus. Some species of phacelia are quite localized in their distribution. This is true of *Phacelia idahoensis,* for the plant is found only in the central counties of Idaho (this plant was photographed in Custer County, west of Stanley). The ending *-ensis* used with a species name means "originating in."

Thread-leaf phacelia
Phacelia linearis (Pursh) Holz.

The thread-leaf phacelia bears showy, pink to light purple flowers. Typically it is found in foothills and at lower elevations in the mountains. It is a rather common late spring to early summer plant that favors dry surroundings where its hairy leaves serve to conserve moisture. The leaves, while narrow, hardly seem narrow enough to deserve the name "thread-leaf." Unlike the Idaho phacelia, *Phacelia linearis* is common in the western provinces and states, east of the Cascade and Sierra Nevada ranges. Meriwether Lewis collected this plant, then new to science, on the expedition's return trip, at today's The Dalles, Oregon, on April 17, 1806.

Dwarf hesperochiron
Hesperochiron pumilus
(Douglas ex Griseb.) Porter

The dwarf hesperochiron is a pretty little plant, with white, loosely clustered, dark-veined flowers. It is so unlike other members of the waterleaf family that when first seen, often in the company of springbeauties (*Claytonia* spp.), one may have trouble identifying it. The plants bloom in the spring on ground still moist from the snowmelt. The word *Hesperochiron* is derived from *hesperius* ("western") and *Chiron*, the name of a mythological centaur. The significance of the name is unknown; possibly it was attached to a similar plant in antiquity. The species name, *pumilus,* means "dwarf" in Latin. The plant grows in all the western states.

Ballhead waterleaf
Hydrophyllum capitatum
Douglas ex Benth.

The ballhead waterleaf is an early spring blooming plant found along seasonal freshets or on slopes still moist from the snowmelt. Attractive, round, frizzy, purple flowerheads up to two inches in diameter soon appear, partially hidden by the plant's bright green, incised leaves. The small flowers have five sepals, five petals and five projecting anthers. Typically purple, the flowers may range to white at subalpine elevations. There are several varieties of the waterleaf distinguished chiefly by the length of the leaf and flower stems. The plants are found in the four states of the Northwest, south to Colorado and Utah and north to British Columbia and Alberta.

Mint Family (Lamiaceae)

Lamiaceae is the scientific name of the Mint family (an older family name, Labiatae, is also correct). The family consists of approximately 265 genera and 6,000 species. Members typically have square stems and many produce volatile oils that are responsible for their strong "minty" odor. Characteristically, their flowers are borne in whorls (verticillasters) above paired, usually toothed, leaves. The flowers are bilaterally symmetrical with five petals, two of which form an upper "lip," explaining the older family name Labiatae (the Latin word *labiatus* means "lip-shaped"). While not all species share the above characteristics, there usually is little problem in identifying members of the family. It comprises many garden plants including species of ornamental *Salvia*, bee-balms (*Monarda* spp.) and others, as well as many useful culinary herbs including various mints (*Mentha* spp.) such as peppermint, spearmint and catnip; sages (*Salvia* spp.); species of thyme (*Thymus* spp.), oregano and marjoram (*Oreganum* spp.), basil (*Ocimum*), rosemary (*Rosemarinum*), lavender (*Lavandula* spp.), California's yerba buena (*Satureja douglasii* Benth.) and the list goes on. About two dozen genera are native to our Northwest including the fragrant native field mint, *Mentha arvense* L., often found in Idaho where it grows in moist places at lower elevations. Several Eurasian members of the family are also quite at home in our Northwest.

Snapdragon skullcap
Scutellaria antirrhinoides Benth.

Although scutellarias do not smell minty and do not resemble garden mint, their irregular, bilaterally symmetrical flowers and square stems suggest that they are in the mint family. This plant, a perennial, is found in Idaho and west to Oregon and Washington, growing on rocky ground to subalpine elevations. A look-alike, somewhat larger plant, the narrow-leaved skullcap, *Scutellaria angustifolia* Pursh (not shown), also is native to the Northwest, although it grows at lower elevations. The common name "skullcap" is said to have been derived from the shape of the calyx, because of its resemblance to a visored helmet. The same prominent bump on the calyx was also interpreted as mimicking a tray or dish (Latin *scutella*) giving the plant its generic name. The species name, *antirrhinoides,* means "snapdragon-like" for its resemblance to that flower, although the two plants are not related.

Western horsemint
Agastache urticifolia (Benth.) Kuntze

The western horsemint (or giant hyssop) grows in great numbers along our mountain trails, especially in areas that retain moisture from the snowmelt. It occurs in all the states contiguous to Idaho, as well as in California and Colorado, growing as high as the subalpine zone. The tall, square-stemmed plants have a pronounced minty odor. *Agastache,* derived from two Greek words, implies "a spike of wheat," reflecting the form of the flowerheads, which range in color from pale to deep purple. The Latin-derived species name, *urticifolia,* means "nettle-leaved" for the resemblance of the plant's leaves to those of the common stinging nettle (*Urtica dioica*).

American dragonhead
Dracocephalum parviflorum Nutt

The dragonhead is encountered occasionally, usually in moist forest surroundings where it grows as high as the upper montane zone. It is considered a weed for it has no known uses, although it is seldom found growing in large enough numbers to be considered noxious. The odorless plant is found in the American West from Alaska to Arizona, across the continent in all our northern states and in the Canadian provinces to Quebec. It is not hard to identify with its jagged leaves and, when mature, flowerheads that are dense with many small, white, bilaterally symmetrical flowers.

Flax Family (Linaceae)

The Flax family is not a large one. It consists of eight genera and about 250 species. It is represented in our mountains by two native plants: the widely distributed Lewis's wild blue flax and King's flax, *Linum kingii* S. Watson (not shown), found only in the southeastern corner of the state. Common, or domestic flax, *Linum usitatissimum*, also may grow wild near settled areas as a garden escapee. Both Lewis's flax and cultivated flax are closely related to the Eurasian wild flax, *Linum perenne*, the presumed ancestor of the domestic plant. As an indication of how closely related the latter three plants are, some classify our wild flax, *Linum lewisii*, as a subspecies of the Eurasian plant. Flax has been an important source of fiber and oil since prehistoric times, evidenced by remnants of linen found in Swiss lake dwellings and in ancient Egyptian tombs. Flax is grown commercially today, both for its fiber (although linen has been largely replaced by synthetic fibers) and for its seed. Flaxseed (or linseed) is about 40 percent oil by weight; it is of great importance to the paint industry. Additionally, flaxseed and flaxseed oil are being used increasingly as dietary supplements and in health foods.

Wild blue flax
Linum lewisii Pursh var. *alpicola* Jepson

The wild blue flax (also known as Lewis's blue flax or prairie flax), *Linum lewisii*, is widely distributed west of the Mississippi River and north to Alaska. Two varieties are recognized. The variety shown here occurs in circumscribed islands in the mountains of central Idaho as well as in Nevada and California. The plants grow on rocky, south-facing slopes, blooming from June to August, according to elevation—we have seen them in a reduced and prostrate form well above treeline. The flowers of this variety are a grayish-blue, unlike the deep blue of plants (var. *lewisii*) that grow at lower elevations. Their name honors Meriwether Lewis, who collected this previously unknown plant on Montana's Sun River on the expedition's return journey (July 9, 1806).

Cultivated flax
Linum usitatissimum L.

Cultivated, or garden flax, is commonly planted as an ornamental, both in gardens and as a roadside plant where it often forms crowded masses of bright blue flowers. These are annual plants—unlike the perennial wild blue flax—that reseed themselves and may persist for years in the same location. There are other differences between the wild and cultivated plants. The wild flax has slightly wider leaves and the alpine variety described above is lighter in color. Also, the cultivated flax is dimorphic in that some flowers have anthers that are longer than the styles, and in other flowers the opposite is true. The species name, *usitatissimum*, means "most useful." Forms of this plant are cultivated both for fiber and for linseed (flaxseed) oil.

Blazing Star Family (Loasaceae)

The Blazing Star family is small, made up of twenty genera and about 320 species. Most are native to the Americas although members also occur in Africa. It is represented in Idaho by species of *Mentzelia* that grow to the montane zone or higher in our mountains. The family has little economic importance other than the occasional use of some of the plants as cultivated ornamentals. As with various evening primroses (Onagraceae), the flowers open in the evening, are pollinated by night-flying insects, and then fade during the day. The family name is taken from a tropical genus, *Loasa*, which includes several ornamental species.

Blazing star
Mentzelia laevicaulis
(Douglas ex Hook.) Torr. & A. Gray

The blazing star is a relatively tall plant with rough foliage. Five of its many stamens are flattened and petaloid, but not nearly as large as its five true petals. The plants are often seen growing on gravelly road embankments, as high as the subalpine zone. It grows in Montana, Wyoming and Colorado, west to British Columbia and the three coastal states. Because the flowers of these two mentzelias are showy, they are sometimes cultivated as garden ornamentals in dry environments. The genus is named for German botanist Christian Mentzel (1622–1701).

Ten-petaled blazing star
Mentzelia decapetala
(Pursh) Urban & Gilg ex Gilg

The ten-petaled blazing star is not easily missed. It blooms late in the day, is moth-pollinated at night, and closes early in the morning; shown here is a flower from a day or so earlier and this morning's flower. During the rest of the day, the plants look like tall, rough-leaved weeds. The ten-petaled flowers include five true petals and five inner modified stamens (staminodes). Lewis and Clark collected this new-to-science plant in August 1804, near today's Homer, Nebraska. Primarily a Great Plains plant, it grows from the Mississippi River west to Idaho where it occurs in scattered locations, growing on south-facing gravelly slopes to fairly high elevations.

Whitestem blazing star
Mentzelia albicaulis (Doug. ex Hook.) Doug. ex Torr

When it comes to botanical names, a suggestion is often as good as a fact, as with this plant whose stems are rather lighter in color compared with its foliage, thus gaining it the name "whitestem" (*albicaulis*). The plant is not uncommon and is found in most of our western states from Montana to Texas, west to the coastal states and British Columbia. Despite its ubiquity, it is not a well-known wildflower, and it is often confused with similar yellow-flowered plants such as potentillas (Rosaceae).

Nevada blazing star
Mentzelia dispersa S. Watson

About thirty genera of blazing stars are found in the United States; only a few of these occur in Idaho, growing to mid-elevations. They are rather inconspicuous plants, like the ones shown on this page, especially when they are compared with the mentzelias on the preceding page. The Nevada blazing star has much the same distribution as the whitestem mentzelia shown above.

Mallow Family (Malvaceae)

The Mallow family is a moderately large one of approximately 197 genera and 2,850 species worldwide. It takes its name from the Latin word *malva*, used in the past for various mallows. The family has considerable economic importance. *Gossypium* species include cotton plants, important for their textile fiber and for oil extracted from their seeds. Species of hibiscus and the related rose-of-sharon (*Hibiscus* spp.) are ornamental garden plants. Okra (*Abelmoschus esculentus* (L) Moench) is also a Malvaceae. The hollyhock (*Alcea rosea* L.) is another popular garden mallow. Most mallows are easily recognizable. Their flowers are five-petaled and grow in terminal clusters. Staminate tubes formed by fusion of the filaments of the anthers protrude from the center of the flowers (very noticeable in the hibiscus, but present in all) and help with identification.

Gooseberry leaf globemallow
Sphaeralcea grossulariifolia (Hook. & Arn.) Rydb.

This plant is sometimes considered a weed because of its casual growth along trails and roadsides. If one believes that weeds are competitive, harmful plants, then this plant is getting bad press. It is non-aggressive and its five-petaled bright orange flowers brighten the landscape. The plant grows in Rocky Mountain and Pacific coastal states. The Greek *sphaera* means "globe" and *alcea* is the name for the hollyhock. The species name, *grossulariifolia,* is also from the Latin and means "gooseberry leaf," from the similarity of this plant's leaves (as shown) to those of the common gooseberry.

Streambank globemallow
Iliamna rivularis (Douglas ex Hook.) Greene

This mallow is sometimes referred to as a wild or mountain hollyhock although hollyhocks are a species of *Alcea*, a genus that does not grow in our area. The name *Iliamna* apparently came from an Athabaskan word for Lake Iliamna in Alaska; *rivularis* means "of brooklets." The latter is an apt term, for this mountain plant blooms in midsummer along streams and in dry creek beds. Its showy pink flowers, the flowers' staminate tubes, and alternating maple-like leaves identify the plant. The streambank globemallow is found mostly in the western mountain states.

Oregon checkermallow
Sidalcea oregona (Nutt. ex Torr. & A. Gray) A. Gray

Many flowers in the mallow family are quite showy. This plant, with its light blue flowers, is common in Idaho. It is tall with lightly veined flowers borne in spike-like clusters. Its alternate leaves are made up of deeply dissected narrow leaflets. The species is quite variable and half a dozen or more varieties are recognized. *Sidalcea oregona* is found in Idaho and surrounding states (although rare in Montana and British Columbia) and in California. It is often seen in moist meadows and with sagebrush, blooming in midsummer.

Cheeseweed, *Malva parviflora* L.

The common cheeseweed and the closely related and very similar dwarf mallow, *Malva neglecta* Wallr. (not shown), are common Eurasian weeds now found throughout the United States and Canada growing to fairly high elevations in our mountains. They are easily identified by their wrinkled round leaves and white flowers with pale purple markings. Their fruit resembles small round cheeses, whence the common name "cheeseweed." Both the fruit and leaves are edible, and are used for food in other countries.

Common mallow, *Malva sylvestris* L.

True mallows—species of *Malva*—are Eurasian plants. Several species arrived in the New World centuries ago and are now at home throughout North America. The common mallow would certainly place high on anyone's list of showy weeds; it resembles a small hollyhock with attractive, purple-streaked, five-petaled flowers. While Americans do not recognize it as a food plant, its leaves have long been used as greens, both raw and cooked. More common in the Pacific coastal states, this plant is now occasionally found at higher elevations in Idaho.

Water-lily Family (Nymphaeaceae)

The Water-lily family is small, consisting of only six genera and about sixty-five species. It is a primitive family of considerable taxonomic interest, as its members are believed to represent relics of early plant forms that preceded the development of monocotyledonous plants and then persisted by adapting successfully to a fully aquatic existence. Various species and their hybrids have been grown for millennia as ornamentals. This is the family's chief economic importance, although the seeds of several species are used elsewhere for food; those of some are said to have narcotic properties. South American species of *Victoria* (named in honor of the English monarch), noted for their enormous floating leaves and fragrant flowers, are the most spectacular members of the family.

Rocky Mountain pond-lily, *Nuphar polysepala* Engelm.

The Rocky Mountain pond-lily is a wide-ranging plant that grows in quiet ponds well up into the mountains. As with other members of the family, the pond-lilies have bulky roots, stems with specialized air pockets, and large fruiting bodies whose seeds were used as food by Native Americans. The thick yellow sepals, small hidden petals and ball-like shape of the flower are typical of the genus. Members are sometimes used as easily grown ornamentals in water gardens. This species grows as far north as Alaska and the Northwest Territories of Canada, and south to California, Arizona and New Mexico.

Evening-primrose Family (Onagraceae)

The Evening-primrose family consists of seventeen genera and 650 species. Although representatives are found worldwide, they are especially abundant in the Americas. Plants from several genera are cultivated as garden ornamentals; examples include North American species of *Clarkia, Epilobium, Gaura* and *Ludwigia,* as well as cultivars of *Fuchsia* from Central and South America. The family's flowers are radially symmetrical with four petals, four sepals and four, or occasionally eight, stamens. In many species, the petals and sepals are joined into a long narrow tube that looks more like a stem than part of the flower. The Onagraceae are well represented in our mountains, especially by *Epilobium* species at higher elevations. The common name, "evening-primrose family," reflects the tendency of various short-blooming flowers in this family to open in the afternoon and fade away the following day, an understandable trait when one learns that they depend on night-flying moths for pollination. It is interesting that many of the species that flower only for an evening open suddenly, often in a matter of seconds. Most bear long "strings" of pollen that are picked up by nectar-gathering moths and carried from one flower to the next.

Spreading groundsmoke, *Gayophytum diffusum* Torr. & Gray

Gayophytum, or groundsmoke, is a spreading, many-branched plant with small, white, four-petaled flowers similar to those of many other members of the evening-primrose family. If one examines the little flowers (best seen with a hand lens), it is easy to see that the flower's "stem" is thick; in actuality, it is made up of the flower's ovary. The sequence shown in the images above is typical for the Onagraceae. White or pink flowers surmount the two-chambered ovary. The flower parts fall away as the fruiting capsule matures (center). The capsules split open to release seeds borne on hairy miniature parachutes (right). Groundsmoke (there are eight species in the genus) are found throughout the West from British Columbia to Arizona and west to the coastal states. The name *Gayophytum* honors French polymath Claude Gay (1800–1873), who traveled to and studied all things Chilean, including its plant life. This plant's species name, *diffusum,* describes its ground-covering growth habit.

Fireweed, *Chamerion angustifolium* (L.) Holub var. *canescens* (Alph. Wood) N. H. Holmgren & P. K. Holmgren.

Because the common fireweed (above left) prefers disturbed ground, it is considered a weed—an undeserved reputation. It spreads by underground stems, reclaiming ground after wildfires (center). As reforestation proceeds, fireweeds are replaced by other flora. Its flowers are attractive with four deep pink petals borne on a long, stem-like flower tube. The latter contains the ovary that matures into a long, pod-like seed capsule giving off wind-borne seeds (right). The two species shown on this page grow all across North America (except for several southern states) and also in Eurasia. They have recently been reclassified from genus *Epilobium* to *Chamerion*, a pre-Linnaean name for the fireweed (or rosebay willow-weed, as it is known in Europe). Our plants are a hairy variety that is present in most of North America.

Red willow-herb, *Chamerion latifolium* (L.) Holub

Willow-herbs are named for the close resemblance of their leaves to those of willows. The willow-herbs shown here are summer-blooming plants that grow as high as the subalpine zone. The red, or dwarf, willow-herb (left) is found only on the sandy banks and sandbars of fast-moving mountain streams. It is a lovely flower whose four red petals alternate with color-matched, lanceolate sepals. The species name, *latifolium,* means "wide leaf," although the leaves are wide only in comparison to those of the closely related fireweed shown above. The name *Chamerion* was derived from two Greek words that meant "dwarf oleander" and somehow became attached to these plants in pre-Linnaean times.

Ragged robin
Clarkia pulchella Pursh

The ragged robin (known also by several other common names including the botanically preferred "pink-fairies") is a plant of the Northwest*, best seen growing wild. The plant is pretty much localized to where Meriwether Lewis collected it near today's Kamiah, Idaho, on June 1, 1806. It was, as Lewis recognized, an unusual plant, unknown to science. It is interesting both for its appearance, and also because it is the only plant of those that the expedition collected that bears William Clark's name. This plant defined the genus *Clarkia.* Now about forty species of *Clarkia* are recognized.

---

*Oddly, *Clarkia pulchella* has also been found (possibly introduced) in separate locations including Ohio, Massachusetts and Vermont.

Rock-fringe
Epilobium obcordatum A. Gray

Finding unexpected blooms in inhospitable places is one of the alpine hiker's great pleasures. So it is with the rock-fringe (also known as the rose willow-herb), a showy, bright pink plant that flowers in July and into August. While native to mountains in Oregon, California, Nevada and Idaho, it is quite localized in its growth preference. The rock-fringe is a subalpine plant that grows in rocky crevices or at the base of south-facing rocks and cliffs. The plants are small-leaved creepers. Its flowers are large—up to two inches in diameter—with four bright pink, heart-shaped (*obcordatum*) petals, eight stamens and a long, dependent style tipped with a cross-shaped stigma.

The generic name, *Epilobium,* was derived from the long, stem-like flower tube that contains the ovary (*epi-* means "on" and *-lobos* refers to the ovary), visible in these flowers and those of the epilobiums shown on the following page.

> **Epilobium Species (*Epilobium* spp.)**
> We can show only a few of the many species of *Epilobium* that grow in our mountains. Any small white or pink flowers with four notched petals, borne on stout flower stems, ranging in size from less than an eighth of an inch up to an inch or more most likely belong to this genus.

Autumn willow-weed, *Epilobium brachycarpum* C. Presl

The autumn willow-weed is a mid- to high-altitude plant that blooms in midsummer. It is a common, spindly, windblown plant that mostly grows on dry ground. Many tiny pink to white flowers have four deeply notched petals lined with reddish veins. Formerly classified as *Epilobium paniculatum* Nutt. ex Torr. & A. Gray, its new species name, *brachycarpum,* means "short fruit." It is widely distributed throughout our West and Canada.

Alpine willow-weed, *Epilobium lactiflorum* Hausskn.

The four-petaled flowers of the alpine willow-weed are tiny, as is the plant overall. Until recently its species name was *Epilobium alpinum* L., but because of confusion with other species of the same name, it has been reclassified as *Epilobium lactiflorum*. Its species name is derived from the Latin *lacteus* for milky and *-florus* for flowered. Both the scientific and common names are misleading, however, for this plant's flowers range in color from pure white through white with pink veins (as in the one shown here), to fully pink or even rose. It is wide-ranging, found throughout the western United States, and north to Alaska. It also grows in Maine, in the northeastern Canadian provinces and east to Greenland and Eurasia. Any small white to pink, high-ranging *Epilobium* with slightly serrated leaves will most likely be this plant.

Common willow-weed, *Epilobium ciliatum* Raf. var. *glandulosum* (Lehm.) Dorn

The common willow-weed has clusters of tiny (less than one-quarter inch), bell-shaped flowers with four deeply notched petals. It blooms from July to mid-August along mountain streams and in other moist situations. Its stems and foliage feel sticky, hence the name *glandulosum* (a "gland," botanically implies a secreting cell with sticky secretions). Note the long tubular "stem" that houses the flower's ovary; this matures into a pod-like seed capsule. The plant is found throughout the West, north to Alaska, in all of Canada and in most of our northern states.

Northern suncup, *Oenothera subacaulis* (Pursh) P. H. Raven

The northern suncup (also known as long-leaf evening prim-rose) is a low-growing, four-petaled summer plant found only on moist ground. The species name, *subacaulis*, means "not much of a stem," referring to the flowers, for the "stem" is part of the flower tube and the actual stem is very short. Its wide leaves are said to be edible greens. The plant ranges from the northern Rockies, west to central Washington and eastern California. It was first collected by Meriwether Lewis on the Weippe Prairie in northern Idaho on June 14, 1806. The generic name has varied back and forth between *Oenothera* and *Camissonia*; the former is now preferred.

Tansy-leaved suncup, *Oenothera tanacetifolia* (Torr. & A. Gray) P. H. Raven

The tansy-leaved suncup's growth habit is similar to that of the plant shown above and it has a similar range, although, in Idaho at least, it is less often seen. It is characterized by pinnate leaves that are rather like those of the common tansy (*Tanacetum vulgare*), explaining both its common and scientific species names. As with the northern suncup, there is confusion as to whether this plant should be placed in genus *Camissonia* or *Oenothera*. Both names are used and may be found in various field guides.

Pale evening primrose, *Oenothera pallida* Lindl.

The pale evening primrose may grow quite high in the mountains, but no matter the elevation, it will be found growing on sandy soil. It is characterized by that growth habitat, by its overall leafy appearance and by its striped, unopened, reddish-brown buds. It grows throughout the West, from British Columbia to Texas and all the states in between, somehow missing California.

Common evening-primrose
Oenothera villosa Thunb. var. *strigosa* (Rydb.) Dorn

The common (or hairy) evening-primrose is found throughout the United States and Canada, except for a few southern states. Several varieties are recognized; ours is a western plant found mostly in the Rocky Mountains (the species name, *villosa,* and the variety name, *strigosa,* both imply that the plant is covered with fine hairs) growing at least as high as the montane zone. Like the rock-rose shown below, this plant has a showy flower whose blooms last but a day. It is usually found in open meadows where it is easily identified by its alternate lanceolate leaves, four large yellow petals and reflexed (bent downward) sepals.

Rock-rose
Oenothera caespitosa Nutt.
var. *caespitosa*

The rock-rose, known by many other local names, is a wide-ranging western plant. It flowers late in the summer and is often found high in our mountains growing on exposed, dry, sandy slopes. It is also seen growing on banked roadsides. Oenotheras bloom at night when moth-pollinators are active, and by the next afternoon the flowers have wilted—today's and yesterday's flowers are present in the illustration. As you hike you had best photograph this showy flower on your way up, as it will be past its prime on the descent!

Peony Family (Paeoniaceae)

The peony family is a small one consisting of only one genus and thirty-three species. Typically white-petaled, most are perennial shrubs or small trees whose alternate leaves are divided into three lobes; each lobe, in turn, divides into three or more smaller lobes. The flowers are usually large and radially symmetrical, with five sepals, five petals (or occasionally ten) and many stamens. Five large seed capsules form while the plant is still flowering, and each contains many seeds. The family's chief commercial value is for its cultivars—well-known ornamental garden varieties with showy blossoms. Peonies were named by Theophrastus (372–c.287 BC), Aristotle's pupil and the author of an important work on botany. The name honors Apollo, who, in his role as Paean, physician to the gods, used the plant medicinally. Certain species of peony are still used in folk medicine (e.g., the European alpine plant *Paeonia officianalis* and an Asian tree, *Paeonia suffruticosa*), although they seem to have no proven therapeutic worth. Only two peonies are native to North America. The western peony, *Paeonia brownii*, shown below, grows in all the northwestern states and south to California, Nevada and Utah. The similar California peony, *Paeonia californica*, is found only in that state.

Western peony, *Paeonia brownii* Douglas ex Hook.

One might not relate the western peony's flowers to those of the popular garden plant; nevertheless they are close cousins. Our plants grow as high as the subalpine zone and are often seen along hiking trails, in canyon bottoms and on nearby sagebrush slopes, blooming in mid- to late spring. The large flowers nod and may be hard to see at first glance, although the plant's light green, incised leaves stand out against the surrounding sagebrush. The flower is so unusual that it is easily identified. David Douglas collected the plant in the mountains of today's Oregon in 1826. Douglas suggested that it be named for Robert Brown (1773–1858), one of England's most prominent botanists. William Jackson Hooker described it several years later in his *Flora Boreali-Americana* (1829) and followed Douglas's suggestion in naming the plant.

Phlox Family (Polemoniaceae)

The Phlox family consists of twenty genera and 360 species. Most are found in North America although a few are native to temperate parts of South America and Eurasia. Many species in the family have been cultivated as garden ornamentals: phlox, Jacob's ladder, gilias and others that represent the family's only economic importance. Flowers in this family are usually clustered and sometimes form a head in which the upper flowers are the first to bloom (botanically, a cyme). Five sepals and five petals unite at the base to form a flower tube. At the point that the petals become separate they flare outward, and the flowers then are variously referred to as being trumpet-, funnel-, or saucer-shaped (the latter are salverform). Fourteen genera of Polemoniaceae occur in the Northwest. At least half of them are represented in Idaho. The plants in some of the genera are so small that they would easily escape notice unless one is looking for them. The word "Polemoniaceae" is derived from "polemonium," the name of a European alpine wildflower, *Polemonium caeruleum* L., commonly known as Jacob's ladder or Greek valerian, a plant that has been cultivated as an ornamental for centuries. The origin of the word "polemonium," in turn, is uncertain; several derivations have been suggested. It may be that the plant was named for Polemon, a second-century BC Greek philosopher.

Mountain navarretia, *Navarretia divaricata* (Torr.) Greene

Mountain navarretia (plants in this genus are also, for obvious reasons, known as pincushion plants) are found high in our mountains. The species name, *divaricata*, means "branching," chosen because, uniquely, stem branches are given off from within the terminal leaf clusters. The plants appear suddenly in moist places. As the ground dries, they become powder-puffs, like ghosts of their former selves. The plant is found in the four northwestern states and west to Nevada and California.

Brewer's navarretia, *Navarretia breweri* (A. Gray) Greene

One or more flowers, surrounded by needle-shaped leaves, grow at the end of this plant's stems. A long flower tube and five-petaled flowers place it in the phlox family. Genus *Navarretia* was named for Francisco Fernandez Navarrete (d. 1689), a Spanish missionary, physician and botanist. William Henry Brewer (1828–1910), for whom this species was named, took part in a survey of California in the 1860s. He later became professor of agriculture at Yale (1864–1903). The plant grows throughout the Rockies and west to the coastal states.

Spreading phlox
Phlox diffusa Benth.
var. *longistylis* (Wherry) M. E. Peck

The word *phlox* ("flame" in ancient Greek) was the name of a now unknown plant. Later, it was given to our flowers. Despite their worldwide popularity as ornamentals, all phlox cultivars seem to have been derived from North American plants. About eighteen species grow in the Northwest; this one, *Phlox diffusa*, is common in our mountains, growing in clumps of narrow gray-green leaves surmounted by a profusion of usually white flowers, although pink and lavender forms are common. Spreading phlox dots rocky slopes from early April at mid-elevations well into the summer at timberline and higher. The species occurs in all our western states and provinces. A look-alike species, *Phlox hoodii* Richardson, is common at lower altitudes.

Longleaf phlox
Phlox longifolia Nutt.

The longleaf phlox grows at all elevations, blooming from late in the spring well into the summer, depending, as always, on the altitude. The plant is taller and characterized by narrower petals, longer leaves and less tightly clustered flowers than those of the spreading phlox shown above; the two plants have a similar distribution. Nathaniel Wyeth (he of the wyethias) collected the original specimens in 1833 in Idaho or Montana, and his friend Thomas Nuttall described the species the following year.

Showy phlox
Phlox speciosa Pursh

The showy phlox grows at lower elevations than do those shown above. It often grows in the company of sagebrush in the foothills and lower mountain ranges. It is more common in the northern half of Idaho. Lewis and Clark collected the plant on May 7, 1806, while in what today is Nez Perce County on their return journey. Showy phlox is found from British Columbia south to California's Sierra Nevada, into Nevada and east to Montana. The plant is easily identified, for this is the only phlox in our area with prominently notched petals. An attractive plant, the showy phlox makes a good garden ornamental.

Slender phlox, *Phlox gracilis* (Hook.) Greene

The slender phlox (formerly *Microsteris gracilis*) appears in the spring, sometimes in large numbers. Its flowers may be pink or white; pink is most common. The flowers are usually borne in pairs, although they do not always bloom at the same time. Typically the petals are notched at the end. Elliptical, opposed leaves become narrower and more pointed toward the top of the stem. The plant is only about two inches high—the smallest of our phloxes—so single plants are easily missed. It is widespread throughout the West, growing as far north as Alaska. The species name, *gracilis*, means "graceful" or "slender."

Diffuse collomia, *Collomia tenella* Gray

Collomias bloom in late spring to early summer. This species is notable for its peculiar involucrum with its black markings and small "horned" projections (inset). The plants are covered with fine hairs, best seen with magnification. Many of the hairs end in tiny pigmented glands that sometimes cause part or even all the plant to appear black. The species grows in all of the states contiguous to Idaho except Montana. Its species name, *tenella*, means "delicate." Recently, in an effort to promote some uniformity for common names, "mountain trumpet" has been suggested as a common name for all the collomias.

Narrow-leaf collomia, *Collomia linearis* Nutt.

The narrow-leaf seems to be the most common collomia. It is widely distributed, occurring in all Canadian provinces and all but a few southern states. Its five-petaled flowers and long flower tube are typical for the family. The name *linearis* describes the plant's narrow leaves, and *collomia*, from the Greek, means "glue," because the seeds become mucilaginous when wet, helping to identify plants in this family. Lewis and Clark collected this plant on April 17, 1806, at today's The Dalles, Oregon, on their homeward journey. Frederick Pursh, who described the expedition's plants, did not recognize the specimen as a new species so the explorers received no credit for finding it; it remained for Thomas Nuttall to find again.

Staining collomia
Collomia tinctoria Kellogg

Evidently this little mountain plant contains a yellow dye—although who would think to check the plant for that property? It grows at high elevations in the Pacific coastal states and in Idaho, Nevada, Montana and Utah. While the inland plants have white flowers, those found in the coastal states are pink. The staining property (found in the roots) has been reported to make this a human-use dye plant, but I can find no reference to it actually having been used for this purpose.

Scarlet gilia, *Ipomopsis aggregata* (Pursh) V. Grant

The scarlet gilia, formerly *Gilia aggregata*, blooms from mid-spring through the summer according to the elevation. *Ipomopsis* means "morning glory-like" and *aggregata* refers to its flower clusters. Scarlet gilias grow in western states from British Columbia south to Oklahoma and Texas. Lewis and Clark collected the scarlet gilia, then unknown to science, on June 26, 1806, while on the Lolo Trail. The earlier name, *Gilia*, honored naturalist and clergyman Filippo Luigi Gil (1756–1821), the director of the Vatican observatory. It is now well entrenched as a common name.

Ball-head gilia, *Ipomopsis congesta* (Hook.) V. Grant

Here is another gilia that has been reclassified as an *Ipomopsis*. The species name, *congesta*, reflects the plant's round flowerhead with its many tiny, phlox-like flowers. It grows high in the mountains on dry sandy slopes, blooming in midsummer. The leaves are silvery-green, usually with three linear leaflets, arising from a central branched, woody stem. Nine varieties of this plant are recognized, classified mostly by the shape of the leaves; ours is var. *viridis* (Cronquist) Reveal. The plants, as one variety or another, are found in most states west of the Mississippi.

Nuttall's leptosiphon *Leptosiphon nuttallii* (A. Gray) J. M. Porter & L. A. Johnson

This attractive plant (originally classified as *Linanthastrum* and more recently as *Linanthus*) is a common, summer-blooming wildflower that grows as high as treeline. Typically it forms a discrete clump. Its linear, alternate leaves are so close together that they appear to form separate rosettes on the plant's woody stems. White, yellow-eyed flowers are five-petaled; each has five prominent anthers. At times there are so many flowers as to completely cover the plants. They give off a faint, sweet odor. Thomas Nuttall found the species, then new to science, near Fort Hall in 1834. The name *Leptosiphon* is a recently restored old classification derived from the Greek; it means "slender tube" for the narrow flower tube.

Western Jacob's-ladder, *Polemonium occidentale* Greene

The western Jacob's-ladder is a moist-meadow and streamside plant that grows to be a foot or more high. It is so closely related to an uncommon European alpine, *Polemonium caeruleum,* that some botanists believe ours is a variety of that plant and classify it as such. Our plants grow as high as the subalpine zone. Furry stems are topped with one or more small, five-petaled, bright blue flowers. Prominent yellow anthers on long filaments add to the flower's attractive appearance. The name Jacob's-ladder reflects the plant's feathery compound leaves, common to the genus. These give off a pronounced skunk-like odor when crushed; the smell may be the first indication that polemoniums are growing nearby. The plants grow in all the Rocky Mountain and coastal states (save New Mexico and Arizona), and north to Canada's Yukon Territory. A disjunct population is found in northern Minnesota.

Showy polemonium, *Polemonium pulcherrimum* Hook.

Lewis and Clark collected a species of polemonium on the Lolo Trail on June 27, 1806. Originally it was believed to be *Polemonium occidentale* (shown above). It seems more likely, however, that it was the showy polemonium, a common species along their trail during that time. While the plant shown here—photographed on a shady slope near Lolo Pass—has pale blue flowers, those growing in the open are often bright blue. The species name, *pulcherrimum*, means "most beautiful." Several varieties are recognized. Ours, var. *pulcherrimum*, grows east to Montana, Wyoming and Utah, west to the Pacific coastal states and provinces, and north to Alaska.

Sticky polemonium, *Polemonium viscosum* Nutt.

The sticky polemonium (or Jacob's ladder) grows in Washington, Oregon, Idaho and Alberta, then south to Arizona and New Mexico. It is an attractive, low-growing plant with deep blue flowers and pinnate leaves populated with small, tightly ranked leaflets. When crushed, this plant's leaves give off a strong skunk-like odor—skunk polemonium is another common name. The plants grow in separate clumps and are found quite high—we have seen small specimens growing well above treeline.

Buckwheat Family (Polygonaceae)

The name Polygonaceae is derived from two Greek words: *poly* meaning "many" and *goni* for "joint," a reference to species that have swollen stem-nodes or joints.* The family is made up of fifty-two genera and approximately 1,105 species. Typically, members have smooth-bordered, unlobed leaves and flowerheads made up of tight clusters of small flowers. Varying numbers of sepals (three to six) substitute for petals. In some species an accessory structure known as an ocrea forms a sheath around the main stems at the node where the leaves join it. A few Polygonaceae are commercially important as ornamentals and as food plants, including buckwheat (*Polygonum esculentum*), sorrel (several *Rumex* species) and rhubarb (*Rheum rhaponticum*). The latter—in common with other members of the family—contains oxalic acid responsible for the plant's acidic juice. Other species of rhubarb (e.g., *Rheum officinale*) were used in the past as purgatives, and a few species have found a place in ornamental gardens.

———

*This derivation is debated. Some believe that the *gon-* stem refers to "seeds" rather than "joints." Thus Polygonaceae would imply "many seeds."

Sulphurflower wild buckwheat
Eriogonum umbellatum Torr., var. *ellipticum* (Nutt.) Reveal

Eriogonum is by far the largest genus of Polygonaceae in our mountains. The genus is made up mostly of western North American plants—a sampling is shown on the following page. In the intermountain region, the number of species in the genus is outnumbered only by the number of penstemons (*Penstemon* spp. in the Snapdragon family) and milk vetches (*Astragalus* spp. in the Pea family). As with those plants, identification of wild buckwheats can be difficult, made more so by hybridization and variations in form and color.

The sulphurflower* wild buckwheat is sometimes known as the "umbrella buckwheat," possibly a better name because it reflects the species name *umbellatum* (Latin for "umbrella"), referring to the flowerheads and their stems, reminiscent of the umbels that characterize the Parsley family (Apiaceae). This species is found from the Rocky Mountain states west to the Pacific Coast and north to British Columbia. At last count some forty varieties were recognized. The variety shown here is localized to portions of Washington, Oregon and Idaho and has a highly branched umbellate inflorescence. Another variety, with cream-colored flowers and a simple umbel, var. *dichrocephalum* Gandoger, is also common in our area.

———

*By convention, scientific botanical names are spelled with "ph" and animal names with "f" in words such as "sulfur/sulphur." Either spelling is, of course, correct.

Purple cushion wild buckwheat
Eriogonum ovalifolium Nutt.
var. *purpureum* (Nutt.) Durand

The purple cushion wild buckwheat oc-curs in all western mountain states and provinces. It is characterized by small, oval, gray-green leaves, a tightly formed flowerhead and creamy-white flowers that develop purple coloration as they mature.

Dwarf cushion wild buckwheat
Eriogonum ovalifolium Nutt.
var. *depressum* Blank.

The dwarf cushion wild buckwheat is native to the northern Rocky Mountains, growing at subalpine to alpine elevations. When mature, its flowers are often bright red. The cushion buckwheats shown on this page are two of eleven varieties of *Eriogonum ovalifolium*.

Hitchcock's wild buckwheat
Eriogonum capistratum Reveal
var. *capistratum*

Hitchcock's wild buckwheat is an eye-catching, matted plant made up of tightly clustered, tiny gray leaves and two-tone yellow and red flowers. While this variety occurs in the foothills and lower moun-tains of central Idaho, an alpine variety, var. *muhlickii*, is found in the high moun-tains of western Montana.

Piper's wild buckwheat
Eriogonum flavum Nutt.
var. *piperi* (Greene) M. E. Jones

Many eriogonums have yellow flowers; none are brighter than those of this plant. Its narrow lanceolate leaves are usually smooth on top and hairy beneath. The plants prefer rocky ground at high altitudes. It is found in the mountains of our Northwest and as far north as Alaska.

Parsnip-flower wild buckwheat
Eriogonum heracleoides Nutt.

This eriogonum grows in the northern Rocky Mountains and west to mountains in Oregon, Washington and California. In Idaho it is a montane plant, often the dominant valley wildflower in the dry heat of early summer. A whorl of bracts, mid-stem below the creamy-white flowerhead, identifies the plant.

Hairy Shasta wild buckwheat
Eriogonum pyrolifolium Hook.
var. *coryphaeum* Torr. & A. Gray

A high-altitude plant, this alpine wild buckwheat grows near treeline, where it blooms soon after snowmelt. Identify it by its smooth, bright green basal leaves and white flowerheads. It is native to the high mountains of all four states of the Northwest.

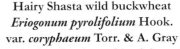

Matted wild buckwheat
***Eriogonum caespitosum* Nutt.**

The word *caespitosum*, from the Latin, reflects the common name, meaning "matted" or "growing in tufts or clumps." The leaves are densely ordered, forming a matted background for the yellow flowers, typical of the genus *Eriogonum* in general. A woody-stemmed perennial, it is easily identified by its dense growth. The plants, like the others shown here, grow at high elevations in our mountains. It is found in Idaho and contiguous states to the south and southwest.

Wicker-stem wild buckwheat
***Eriogonum vimineum*
Dougl. ex Benth.**

Not all eriogonums look like those on the preceding pages. This plant, for example, consists of supple, rambling stems with flowers attached by dainty stemlets. It is as if the plant had been expanded to show off its small clusters of typical eriogonum-like flowers. The name *vimineum* means "osier-like" (i.e. stems suitable for basket making). The plant is restricted to Idaho, Nevada and the three Pacific coastal states.

Alpine bistort
***Polygonum viviparum* L.**

The alpine bistort has been variously classified (*Polygonum viviparum, Persicaria bistorta*, etc.), reflecting the unusual way in which it spreads. A thin spike bears many stemless buds. Those at the top mature into white flowers with black stigmas. Buds lower down become bulbils and enlarge to give off small leaves. The bulbils are shed and form new plants. The alpine bistort is a circumboreal plant, found in mountain ranges of our western states and north throughout the arctic and mountains of both Old and New Worlds. Its bulbils are rich in starch, and the plant is used making Lent- or dock-pudding in northern England. A large, showy cultivar of the plant is used in ornamental gardens.

American bistort
Bistorta bistortoides Pursh

The American bistort grows in mountain meadows throughout the West. The name is an old one, derived from the Latin *bis* ("twice") and *torta* ("twisted") and applied to a similar Eurasian plant. It refers to the plant's bulky, twisted roots. These store the food required for rapid growth during a short growing season. The American bistort blooms from June well into August. Small white flowers form well-delineated clusters at the top of spindly stems. Both the roots and the young leaves of the bistort are edible and were used as food by Native Americans; the flowers and foliage are grazed by deer, and bears dig up the roots. Lewis and Clark were the first to collect this plant, on the Wieppe Prairie in north-central Idaho on June 12, 1806, on their return journey. Recent genetic studies suggest that this plant deserves to be classified as its own genus, or in the same genus as the Eurasian bistort, rather than as a species of *Polygonum* as the plant was formerly classified.

The larger illustration shows white American bistort interspersed with bright orange Indian paintbrush, growing above treeline at the foot of Old Hyndman Peak, in central Idaho's Pioneer Range.

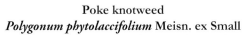

Poke knotweed
Polygonum phytolaccifolium Meisn. ex Small

The poke knotweed, variously known also as the alpine knotweed, pokeweed, or fleeceflower, bears some slight resemblance to the common pokeweed (*Phytolacca americana* in the family Phytolaccaceae), and the species name, *phytolaccifolium*, implies that its leaves resemble pokeweed leaves.* The poke knotweed is a common subalpine plant found in moist situations where it blooms from midsummer on. Given its distinctive appearance, it is unlikely to be confused with any other Polygonaceae. While it is without food value for humans, animals graze freely on the plant.

---

*The etymology of the *lacc-* stem is interesting. It is derived from a Hindustani word, *lakh*, for an insect that secretes a bright red, sticky material used as a pigment—whence the artist's color "crimson lake." The same material, when decolorized and dried is known as "shell lac"—whence "shellac."

Mountain sheep sorrel
Rumex paucifolius Nutt.

The mountain sorrel's clustered flowers turn conspicuously red as they fruit, so the plants form red patches in mountain meadows where they grow. The species name, *paucifolius*, means "few-leaved," pertaining only to the flower-bearing stems, for the plants have many lance-shaped basal leaves. *Rumex* is the Latin word for "sorrel." The genus also includes the edible sorrel, *Rumex acetosa*, whose acidic taste is derived from the oxalic acid common to the leaves of all species of *Rumex*. The plants grow in most of our western states. They are unrelated to wood sorrels (*Oxalidaceae, Oxalis* spp.), whose foliage also contains oxalic acid.

Willow dock
Rumex salicifolius Weinm.
var. *triangulivalvis*
(Danser) J. C. Hickman

Willow dock grows in most of the United States. Six (of seven) varieties are found in the Northwest. The variety shown here is restricted to high elevations in the northern Rocky Mountains. The species name, *salicifolius*, means "willow-like leaves." At first glance, it is not an especially attractive plant, although the flowers and the fruit (achenes, shown here) are rather striking. Willow dock is considered a weed in much of its range, but our plant grows too high to have a weedy impact on anything. The varietal name, *triangulivalvis*, refers to the form of the plant's fruit.

Purslane Family (Portulacaceae)

The Purslane family, Portulacaceae (twenty-eight genera, 440 species), gets its name from an Old World potherb, *Portulaca oleracea*. Pliny, in his *Natural History* (first century AD) named the plant *porcil-aka*, a word with no known meaning that, in time, became portulaca and eventually was incorporated into the family's scientific name. Then, because it sounded like "porcelain" (or "purslane") the flower took that as a common name. "Purslane" has been in use since the fourteenth century for the plant. The common purslane, the only European member of the family, is still grown in Europe as a potherb. Introduced to America, it has spread throughout the United States and is now a common, troublesome, fleshy-leaved garden weed (although Harrington, in *Edible Native Plants of the Rocky Mountains,* extols the edible virtues of this easily identifiable plant). Most members of the Purslane family grow in the temperate zones of the Americas and are well represented in our Northwest. Distinguishing family features include large roots (used by Indians as food), fleshy leaves (edible in some species), showy flowers (some are cultivated as ornamental plants, the family's only commercial importance), four to many petals; and, in most, two sepals—the latter is a distinguishing family feature.

Common bitterroot, *Lewisia rediviva* Pursh

When Frederick Pursh, the German botanist, then residing in Philadelphia, saw Meriwether Lewis's specimen of the plant that later became known as the bitterroot, he recognized that it belonged in a new genus. He gave it the generic name *Lewisia*, honoring the explorer. Lewis had collected his specimen at Traveler's Rest near today's Missoula, Montana, on July 1, 1806 (a site worth visiting today). Pursh gave the plant the species name *rediviva*, meaning "return to life," because one of Lewis's specimens bloomed when planted after the expedition's return. The bitterroot is a common plant that grows as far south as California and Arizona (where it is rare), north to British Columbia, as far east as Montana and Wyoming and west to Washington. The plant's many-petaled flowers are truly striking. They bloom on open, gravelly ground, appearing in late spring. The two plants shown above have much the same distribution, although we and others have noted that the white form grows farther south, in Blaine and Camas Counties in Idaho and south to Nevada. It has been classified recently as a separate variety, var. *minor*. Farther north the flowers range in color from light to dark pink (var. *rediviva*). Indians prized the bitterroot's roots, but white men found them totally unpalatable, explaining the plant's common name.

Alpine bitterroot, *Lewisia pygmaea* (A. Gray) B. L. Robins

The pink-flowered alpine, or pygmy, bitterroot (var. *pygmaea*) grows on bare ground from the subalpine zone to far above treeline. The plants are tiny—those pictured above were less than one-quarter inch in diameter, and no part extended more than an inch above ground level. It is difficult to see how the plant survives, subjected to harsh winds and freezing temperatures, but microclimate is everything. There is little wind at ground level, and rocky tundra, warmed by the sun, radiates warmth. So long as the plant hugs the ground, it survives. The white form on the right also occurs in Idaho and neighboring states. It appears to be var. *nevadensis* (A. Gray) B. L. Robins (some classify this variety as a separate species). The pygmy bitterroot grows in all the western states, although it is rare in Colorado, New Mexico and the Canadian province of Alberta.

Springbeauties and Candyflowers (below and on following page)

Species of *Montia* (candyflowers) and *Claytonia* (springbeauties) are so closely related that most have been cross-classified between the two genera over the years. Consensus now favors the classification used here. The name *Claytonia* honors Virginia botanist John Clayton (1694–1773), who first collected the eastern springbeauty *Claytonia virginica*. More than a dozen species of springbeauty—some quite uncommon—are listed as growing in Idaho.

Lanceleaf springbeauty
Claytonia lanceolata Pursh

The lanceleaf springbeauty bears white-petaled flowers adorned with pink veins and anthers. Some years many of the flowers are pink (near left) for reasons that are obscure. Lanceolate leaves give the plant its species name. There are many varieties—at least eight have been described. Ours, var. *lanceolata,* grows in Idaho and the six contiguous states, as well as in the two Canadian provinces to the north. The lanceleaf is the most common of all the springbeauties, in Idaho at least.

Miner's lettuce
Claytonia perfoliata Donn ex.Willd.

Streambank springbeauty
Claytonia parviflora Dougl. ex Hook.

Heartleaf springbeauty
Claytonia cordifolia S. Watson

Miner's lettuce grows in all the western states and north to Alaska. It is unusual because its stems pierce two opposing conjoined leaves. The plant was known as a salad green in the early days of the West.

Streambank springbeauty is a reclusive plant found in the Rocky Mountain and Pacific coastal states, favoring moist woods and quiet streambanks. It is the least common of the plants on this page.

Heartleaf springbeauty prefers deeply shaded, moist forests. Its species name, *cordifolia*, describes the plant's wide, heart-shaped leaves. It is native to the northern Rocky Mountain and Pacific coastal states.

Chamisso's candyflower
Montia chamissoi (Ledeb. ex Spreng.) Greene

This plant is usually classified as a species of *Montia* (candyflowers) rather than as a *Claytonia* (springbeauties). It is named for a German botanist who discovered the plant. While all spring beauties prefer moisture, this plant grows in the water of spring puddles and freshets, often in large numbers.

Siberian springbeauty
Claytonia sibirica L.

Siberian springbeauties are found in all the Pacific coastal states, north to Alaska and inland to Idaho and Montana. A streamside plant, it is characterized by broad leaves and loose clusters of white flowers borne on long, delicate stems. In common with most of the other plants in this group, it is common to find them along streambanks and in other moist places.

Primrose Family (Primulaceae)

The Primrose family's common name was derived from the term "prime rose," applied to certain flowers—daisies, primroses, and others—that bloom early in the spring. The family consists of eighteen genera and 955 species. Its members typically have radially symmetrical flowers whose five (occasionally four) petals may join for part or sometimes all of their length into a tube. Most are perennial, herbaceous (non-woody) and native to the north temperate zone. Subalpine and alpine Primulaceae are found in all the mountain ranges of the Northern Hemisphere. Many primulas are grown as ornamental cultivars (primroses, shooting stars, cyclamens, etc.); otherwise the family has little commercial importance.

Jeffrey's shooting star, *Dodecatheon jeffreyi* Van Houtte

Jeffrey's shooting star grows on the banks of mountain lakes and in moist meadows. An alpine/subalpine plant, it blooms from late spring into summer. Its flowers have five swept-back, white-based pink petals with a yellow collar at the base. Purple-brown anthers join to form a point from which the style protrudes. The stigma at the end of the style is about twice the width of the style, a distinguishing feature. Although we have retained *Dodecatheon* here as a name for shooting stars, recent studies have shown that they are best classified as *Primula*. Jeffrey's shooting star grows in the coastal states north to Alaska, and in Idaho and Montana.

Many-flowered shooting star
Dodecatheon pulchellum
(Raf.) Merr.

The many-flowered, or dark-throat shooting star is another *Dodecatheon* often encountered in Idaho's mountains. While both this species and Jeffrey's shooting star, shown on the previous page, are western plants, *Dodecatheon pulchellum* is more widely distributed; it is found in most states and provinces west of the Mississippi River. A squiggly purple ring at the base of the petals is a distinguishing characteristic, as is the size of the stigma, which—unlike that of Jeffrey's shooting star—is the same diameter as the style. Shooting stars typically have a rosette of basal leaves; the shape of the leaves varies with the species. The stem is naked, topped with one to several flowers. Botanists recognize several varieties of this species, differentiated chiefly by the color of the anthers. The plant in the illustration, with its pale yellow anthers, is var. *cusickii* (Greene) Reveal. The term *pulchellum,* from the Latin, means "beautiful."

Rocky Mountain androsace
***Androsace montana* A. Gray**

Until recently this plant, also known as a dwarf primrose, was classified as *Douglasia montana*. On the basis of recent studies, however, it is now classified as an androsace (pronounced andros-a-KEY). The Rocky Mountain androsace is a low, mat-forming, vibrant pink-flowered plant. The flowers have a small, ringed, central "eye," common to primroses in general. It is a true alpine plant, found only in the mountains of Idaho, Wyoming, Montana and, rarely, in Alberta. Other species of *Androsace* occur everywhere in North America to the arctic and Greenland, except in a few southern and eastern seaboard states.

Cusick's primrose
Primula cusickiana
(A. Gray) A. Gray

Cusick's primrose is an early spring-blooming plant whose yellow-eyed flowers range from pale to deep purple. It is found only in the Wallowa Mountains of northeastern Oregon and in central Idaho, where it blooms, after snowmelt, on rough, seemingly unpromising ground from foothills to the subalpine zone. The species was named for William Cusick (1842–1922), a rancher, teacher and botanist who collected plants in the Northwest. Attempts to cultivate this elegant little flower have not been successful. Parry's (or brook) primrose, *Primula parryi* A. Gray (not shown), is another subalpine plant in the same genus. It grows in Idaho (where we have not encountered it) and in the other Rocky Mountain states. The plant is larger than Cusick's primrose—its leaves may be ten inches or more in length. The flowers are reddish-purple with yellow eyes. The plant prefers moist surroundings.

Buttercup Family (Ranunculaceae)

The Buttercup family is a moderately large one, made up of about sixty genera and nearly 2,505 species. Because of the flowers' simple configurations and frequent lack of petals, the family is considered to be one of the more primitive of the dicotyledons. While most members are herbaceous (non-woody) plants, a few are woody shrubs or vines. The family favors the north temperate zone and is very well represented in our Northwest. Although it's a surprisingly diverse family, one whose members take many forms, there are common characteristics: flower parts are free and not joined; many members have petal-like sepals and true petals are often lacking; the leaves are usually compound and three-parted; and the plants favor moist environments.

Most of the Ranunculaceae produce poisonous alkaloids. Some of these are therapeutically active although seldom used today, having been supplanted by safer and more effective medications. Aside from the negative impact that ingesting certain of the Ranunculaceae (*Delphinium* spp. especially) has on cattle and sheep, many members of the family have showy blooms and have found a place in ornamental gardens; e.g., species of *Aconitum* (monkshood), *Aquilegia* (columbine), *Clematis*, *Delphinium* (larkspurs), *Ranunculus*, etc.—these represent the family's chief economic importance. The generic name *Ranunculus* (from which the family name is derived) is a diminutive form of the Latin *rana,* the word for "frog" (i.e., something found in wet places).

Red baneberry, *Actaea rubra* (Ait.) Willd.

The baneberry grows in all our northern states as well as in all the Canadian provinces. It is a shrubby plant, commonly seen growing along the banks of shaded streams. Short-lived white flowers form tight clusters, and its compound leaves are made up of several leaflets. The shrub's red (occasionally white) berries are poisonous. A tincture derived from the European baneberry, *Actaea spicata*, was formerly used in folk medicine to treat pulmonary problems. The name *Actaea* came from the Greek word for the (unrelated) European elderberry because the baneberry's leaflets apparently resemble those of this elderberry. The name *rubra,* from the Latin, means "red," referring to the berries. The common name "baneberry" is derived from the Old English word "bane," a word that means "poison."

Western monkshood
Aconitum columbianum Nutt.

This is the only species of monkshood found in the North-west. It flowers in midsummer in wet meadows, seep-springs, and along streambanks at mid- to high elevations. Deep purple flowers are spaced along the top of a tall stem. The "hood" is a petal-like sepal that encloses two small petals. All parts of the plants are poisonous. Criminals were executed and wolves were poisoned with a distillate from A*conitum lycoctonum*, the European wolf's-bane. The same plant was also supposedly a component of witches' brews. Although our plant's flowers, like those of the related larkspur, are usually purple, mixed color or albino forms turn up occasionally (above). The western monkshood grows in British Columbia, the Rocky Mountain states, the Pacific coastal states and east as far as South Dakota and Iowa.

White marsh marigold
Caltha leptosepala DC.

The marsh marigold is an early blooming, subalpine to alpine plant often found in profuse numbers in wet mountain meadows and on the banks of the seasonal ponds that form as snow melts. Its deep green leaves appear first, forming elongated heart shapes. One to several stems, each bearing a showy white flower with a bright yellow center, then appear. The "petals" are actually petaliform sepals that are sometimes lightly tinged with blue. *Caltha*, a Latin word, was used for a common yellow marigold; *leptosepala* means "slender sepals." The plant grows in all the far western states and north to Alaska.

American pasqueflower
***Anemone patens* L.
var. *multifida* Pritz.**

Our pasqueflower is a lovely plant with large flowers and sepals that range in color from light blue to deep purple. Deeply incised leaves are characteristic and distinguish this plant from the similar western pasqueflower, *Anemone occidentalis*, whose leaves are less dissected; it too grows in Idaho. The American pasqueflower is found in all the Rocky Mountain states, east to the Great Lakes and north to Alaska. The name "pasqueflower" has long been used for a European anemone. John Gerard (1545–1612) in his *Herbal* of 1597 wrote, "They flower for the most part about Easter, which hath mooved me to name it Pasque flower, or Easter flower."

Cliff anemone
***Anemone multifida* Poir.**

The cliff anemone grows on moist ground, from mid-elevations to alpine tundra. The species name, *multifida*, means "much divided," referring to its deeply divided leaves. The flowers are apetalous, (i.e, the "petals"—usually five or six—are actually petal-like sepals). These vary greatly in color, from off-white through ochroleucous (pale yellow) to deep red, bluish or even purple. The plant grows in the West and all across the northern part of the continent. Given the amount of variation, it is not surprising to learn that a half-dozen varieties are recognized. The rosette of leaves (bracts) on the stem below the flower is a distinguishing feature of anemones in general.

Small-flowered anemone, *Anemone parviflora* Michx.

The small-flowered, or northern, anemone is usually found growing close to water, in moist meadows and on streambanks, often at subalpine elevations. Each plant bears a single flower made up of petaloid sepals. In the center many yellow stamens surround a spherical head made up of green achenes. This anemone may be confused with other white-blossomed Ranunculaceae, especially the marsh marigold, *Caltha leptosepala* (shown on page 162). Anemone, from the Greek, is said to mean "daughter of the wind," explaining why anemones are also known as "windflowers." Another common name, thimbleweed, reflects the appearance of the fruiting head.

Piper's anemone, *Anemone piperi* Britton ex Rydb.

Piper's anemone is common in the mountains of north-central Idaho, in the adjacent corners of Washington and Oregon and in western Montana. It is characterized by three compound leaves below a single, delicate white flower. It prefers the moist ground of shaded forests. Meriwether Lewis collected this— then unnamed—anemone near the Clearwater River on June 15, 1806. Frederick Pursh, who classified the expedition's material dropped the ball; apparently he concluded that Lewis's specimen was the same plant as an eastern anemone, and did not include it with other expedition specimens in his *Flora* of 1813.

Colorado blue columbine, *Aquilegia caerulea* James

The blue columbine, Colorado's state flower, is one of our most attractive plants. Although not common in our experience, it may be encountered from time to time throughout the state. The blue flower shown here is the variety most commonly seen; at least four other varieties have been described. Its native range includes the four states contiguous to Idaho on the east and south as well as Colorado and New Mexico.*

---

*The Colorado blue columbine has special meaning for one of us (ASE), who, while an eighteen-year-old mountain trooper training in Colorado in 1943, encountered this lovely blue flower growing alone, far above treeline, on the vertical wall of a steep couloir. The encounter remained in memory for a lifetime.

Sitka columbine
Aquilegia formosa Fisch. ex DC.

The Sitka (also known as red, or crimson) columbine is a handsome, gaudy flower. Tall, with distinctive three-parted leaves, its sepals vary in color from pale orange to bright red, depending on growth conditions. These are set off by five red-based, yellow petals and many long, yellow anthers. The plants flower from late spring through midsummer, as high as treeline. Look for it in open forests, along streams and on high rocky slopes soon after the snowpack has melted.

The scientific name *Aquilegia* was derived from the Latin word *acquila* meaning "eagle," because the flower's five spurs were thought to resemble an eagle's claws. The species name, *formosa,* also from the Latin, means "beautiful." Paradoxically— given the derivation of the generic name—"columbine" means "dove-like," because the spurs in some species are said to be shaped like the head and neck of a dove.

Yellow columbine
Aquilegia flavescens S. Watson

The yellow columbine is another montane to alpine species that grows in Idaho's mountains, although it is less common than the Sitka columbine. It is easily identified by its soft yellow sepals ranging at times to a pinkish color. It is closely related to *Aquilegia formosa,* and the two species may form hybrids. All gradations of color between the vivid red and yellow of the former plant and the overall soft yellow of this one are seen from time to time.

Western virgin's bower
Clematis occidentalis (Hornem.) DC.
var. *grosseserrata* (Rydb.) J. S. Pringle

The western virgin's bower is a woody climbing vine of the mountain West. The flowers are apetalous, and its purple sepals stand out in the shade of the montane forests that the plant prefers. Three-parted leaves with toothed, heart-shaped leaflets are a clue that it is in the Buttercup family. A similar plant that differs in the shape of its leaves, *Clematis columbiana* (Nutt.) Torr. & Gray, is found in the southeastern part of Idaho. The Greek word *klematis* referred originally to a periwinkle (*Vinca* sp.) and later was applied to this genus.

Vase-flower, *Clematis hirsutissima* Pursh

This plant, known also as the sugar-bowl, leather-flower and hairy (*hirsutissimma*) clematis, could not be mistaken for any other. Four purple sepals are joined for much of their length to form a furry "vase" that gives the flower one of its common names. It is a herbaceous (non-woody), soft-stemmed, perennial plant, with flowers borne on a single stem that arises from a profusion of leaves that divide into narrow leaflets. Vase-flowers prefer moist meadows, blooming in mid-spring to early summer. The plant was unknown to science until Lewis and Clark collected it on May 27, 1806, near their encampment on the Clearwater River near today's Kamiah, Idaho. It grows in all the Rocky Mountain states as well as in the four northwestern states.

Western clematis, *Clematis ligusticifolia* Nutt.

The western clematis (or western virgin's bower) is native to most of the western states and Canadian provinces, south to northern Mexico. The plants prefer dry, open ground where they form aggressive, rapidly spreading vines that cover neighboring trees, shrubs and fences. They produce masses of white blossoms, followed by densely hairy fruiting bodies (seen also in other members of the genus). In this plant the fruiting bodies often coalesce to cover the plant. The western clematis is sometimes used as an ornamental, but it tends to spread farther and faster than one might wish. The species name, *ligusticifolia,* apparently refers to a perceived similarity between this plant's leaves and those of a species of *Ligusticum* in the Parsley family (Apiaceae).

Upland larkspur
***Delphinium nuttallianum* Pritz. (above)**

The upland larkspur grows as high as treeline in all the Rocky Mountain states. It is often found on over-grazed land. Cattlemen hate the plant for it poisons livestock—a good example of the maxim "my wildflower, your weed." The plants are usually only six to ten inches high, and flower early in the spring on dry ground, often surrounded by sagebrush. Delphinium flowers are made up of five outer sepals that enclose much smaller petals. The upper sepal forms the distinctive spur. This plant's flower color varies from deep blue or purple, through an attractive soft blue-gray (upper right), to white. Regardless of the flower color, some blue marking is always retained on the upper petals.

Slim larkspur
***Delphinium depauperatum* Nutt. (left)**

Several tall larkspurs grow in our mountains; the slim larkspur is one of the more common ones. The species name, *depauperatum* means "impoverished," although it doesn't seem to fit the plant. It may grow as a solitary plant, or as a cluster, in meadows and in the partial shade of open woods. Telling the various *Delphinium* species apart is not always easy, but this one can be identified by its three-lobed basal leaves (not shown) and by the tiny bracts (leaflets) on each of the flower stems. It also blooms later than the upland larkspur and is considerably taller. The slim larkspur grows in the Pacific coastal states and inland to Idaho, Montana and Nevada.

Tower larkspur
Delphinium glaucum S. Watson*

This larkspur is an impressive, often several-stemmed, mountain plant that may grow to be six feet tall or higher. It grows along streambanks where its many bright blue to purple (or, occasionally, almost white) flowers stand out against surrounding green foliage. Also known as the dunce-cap delphinium for its long-spurred flowers, its appearance and preference for streambanks help to identify the plant. By way of confirmation, the stems below the flowering portion are fistulous (hollow), and its large compound leaves are palmate, made up (usually) of five three-lobed leaflets. The leaves in some plants are noticeably glandular (sticky). The tower larkspur is native to all four northwestern states, north into British Columbia and through the Rocky Mountains to New Mexico.

*This plant is usually classified as *Delphinium occidentale* (S. Watson) S. Watson. Present thinking is that *Delphinium occidentale* is a hybrid between this plant, *Delphinium glaucum* which it resembles, and *Delphinium barbeyi* (Huth) Huth (not shown).

Buttercups, *Ranunculus* spp.
Over forty species of buttercups grow in our Northwest, too many to show here. Many of these are at home in the mountains of Idaho. Although simple, bright yellow flowers are thought of as a norm for buttercups, the reader will note, on the next few pages and in the field, how many different forms the flowers take.

Blue Mountain buttercup
Ranunculus oresterus L. D. Benson

The Blue Mountain buttercup grows from the Blue Mountains of northeastern Oregon, eastward into Idaho and along a swathe across the central part of the state to Elmore (and possibly Camas) County. It blooms, often in large numbers, in moist mountain meadows at lower elevations. The plants are easily distinguished from other buttercups by their clustered appearance, each having five or more flowers, and by their linear, grass-like leaves. The plants are described as having five sepals and petals to a flower, but as can be seen in the illustration, there are often more.

Pink buttercup
***Ranunculus andersonii* A. Gray**

Pink buttercups bloom early in the spring. They were first collected by Dr. C. L. Anderson, for whom they were named, near Carson City, Nevada, in 1866. It is hard to believe that the two plants shown here are the same species. On the left is the form that gave the plant its common name. Usually, however, the flowers are pure white when they bloom, as shown on the right. Within a few days, frilly, deeply lobed leaves identical for both forms appear. Why the flowers take on a pinkish hue in some plants and are white in others is not known.

White water buttercup, *Ranunculus aquatilis* L.

The water buttercup (also known as water-crowfoot) has yellow-centered, mostly five-petaled, white flowers whose stems extend above slow-moving water. Three-parted leaves form when the plants are out of water, but when submerged the leaves divide to form soft, hair-like leaflets that offer little resistance to the current. The water buttercup is a circumboreal plant that grows in many western states, north to Canada and Alaska, and in Europe.

Plantain-leaved buttercup
***Ranunculus alismifolius* Geyer ex Benth.**

The plantain-leaved buttercup appears soon after snowmelt at higher elevations. It is identified by the unusual appearance of its flowers with five (usually) tiny petals, and by round leaves that bear some resemblance to those of water-plantain (*Alisma* spp., Alismataceae), explaining both common and species names. It is found in most western mountain states and (rarely) in British Columbia. Several varieties are recognized, identified by the number and size of the petals, the shape of the leaves, etc. Ours is var. *alismellus* A. Gray.

Subalpine buttercup
Ranunculus eschscholtzii Schltdl.
var. *trisectus* (Eastw. ex B. L. Rob.) L. D. Benson

The subalpine, or Eschscholtz's, buttercup is a high-altitude species found throughout the western mountains and north to Alaska. It blooms in moist declivities that catch the snowmelt, often tunneling up through the receding edges of snowfields. The plant is characterized by a thick stem and three-parted leaves. The lobes divide into three segments (not yet developed here). Its flowers have (mostly) five petals and a calyx made up of as many sepals. Several varieties are recognized; ours seems to be var. *trisectus*. Johan Friedrich Eschscholtz (1793–1831) was an Estonian who accompanied Kotzebue's around-the-world-expedition (1815–1818).

Sage buttercup
Ranunculus glaberrimus Hook.
var. *ellipticus* (Greene) Greene

Most buttercups are easy to identify generically, but species classification—based in large part on the appearance of their fruit (achenes)—can be difficult. The sage buttercup is one of the most recognizable because its bright yellow petals have a shiny, waxy gloss; *glaberrimus* means "smoothest." It is a common montane, early spring-blooming plant that favors open sagebrush slopes. The oval leaves of the sage buttercup are often "entire," meaning, they are smooth edged, without lobes or leaflets (although some plants also have three-parted compound leaves). Each plant usually bears several flowers and most have five petals, although the number varies. The sage buttercup is common throughout the West.

Hillside buttercup
Ranunculus jovis A. Nelson

The hillside, or Jove's, buttercup grows on open montane to subalpine sagebrush-covered slopes, blooming soon after snowmelt. Its petals are not prominent, and some flowers seem not to have any petals. Short-stemmed, lanceolate, three-parted leaves help to identify the plant. Jove's buttercup is found in the intermountain states and in Colorado. The type specimen was collected by Rocky Mountain botanists Ruth Ashton Nelson and Aven Nelson. We have not been able to learn to what or whom *jovis* refers—presumably to the Roman god, but why?

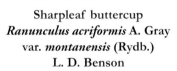

Sharpleaf buttercup
Ranunculus acriformis A. Gray
var. *montanensis* (Rydb.)
L. D. Benson

The sharpleaf buttercup is a relatively common species that grows at mid- to high elevations in our mountains. It is usually seen in moist meadows and along streambanks. The plants may be identified by their long-stemmed, three-parted, compound leaves. Each of the leaflets is further divided into (usually) three slim, pointed lobes. The flowers mostly have five petals, but there are often more. Two varieties are recognized. Those encountered in Idaho grow in the central part of the state, west to Montana and south to Wyoming and Colorado. Another variety, var. *acriformis*, with shorter leaves, grows only in Wyoming and Colorado.

Greater creeping spearwort
Ranunculus flammula L.

The creeping spearwort is a circumboreal plant, known in Europe by that name for at least a millennium. It was not without its usefulness in olden times, for its sap is highly irritating. Mendicants made use of this property to produce open sores whose appearance would, hopefully, engender a sympathetic response that took the form of a loosening of onlookers' purse strings. The plant grows in moist surroundings throughout the American West, Canada and Alaska, and east to Greenland and Eurasia. Creeping spearworts are highly poisonous when ingested by animals, resulting in necrosis of internal parts. Animals thus afflicted are said to be "spearworty." (OED). All in all, it is a plant to observe but otherwise avoid!

Graceful buttercup
Ranunculus inamoenus Greene

A minor botanical mystery is why Edward Lee Greene (1843–1915) named this species *inamoenus,* a Latin word that means "unlovely." The usual explanation is that he was confused and added the "*in-*"suffix in error, for it means "not" or "un-"when he meant to use simply *amoenus,*" a Latin word that means "lovely" or "graceful." He and the plant are usually given the benefit of doubt, so this buttercup with its attractive fan-shaped leaves is commonly referred to today as the "graceful buttercup." It grows on moist ground, appearing after snowmelt at high elevations in the Rocky Mountain states from Canada to Arizona and New Mexico. The leaves are prominent, but usually there are few flowers and occasionally none.

Western meadowrue
Thalictrum occidentale A. Gray

Thalictrums are found throughout the United States and Canada, but this species occurs only in the West. Its leaves are three-lobed. This species is dioecious, with separate male and female plants. Male plants are distinguished by fringe-like dependent anthers (far left). Female plants have wispy pinkish petals (near left). Meadowrues provide ground cover in shady woods and are sometimes grown in shaded gardens. *Thaliktron* was used by Dioscorides for a Greek plant and later became attached to this genus. True rue is an unrelated evergreen shrub, *Ruta graveolens*, with an unpleasant aromatic odor. Our plants have a similar odor, so they became "meadowrues." Western meadowrue was first collected by the explorer Captain John Charles Frémont (1813–1890) in Wyoming in 1843.

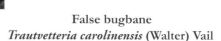

False bugbane
Trautvetteria carolinensis (Walter) Vail

The false bugbane, as its scientific name suggests, is the same annual plant as one that grows in the southeastern United States. It prefers shaded streambanks and moist forests where the ground is often covered with their large maple-leaf-shaped leaves. The flowers are small and gathered into clusters. The generic name, *Trautvetteria*, honors Russian botanist Ernest Rudolf van Trautvetter (1809–1889). Our plant resembles the true bugbane, a related European ornamental, *Cimicifuga foetida* L, used in the past as an insect repellent. The trautvetteria shown here was photographed in the DeVoto Grove, a few miles west of the Lolo Pass summit. The tree is one of the grove's magnificent western red cedars, *Thuja plicata*.

Buckthorn Family (Rhamnaceae)

The Buckthorn family is made up worldwide of fifty-eight genera and 875 species. Most are tropical shrubs or small trees. It is represented in Idaho by plants belonging to two genera, *Ceanothus* and *Frangula*. All the Rhamnaceae have bark that acts as a potent purgative. Other than the occasional use of our plants as ornamental shrubs, the purgative value of *Frangula purshiana* bark (cascara, described below) makes it the only American member of the family to have commercial value. Elsewhere the bark and berries of various buckthorns have supplied dyes used in painting and for textiles. The word *rhamnus,* from which the family name was derived, is an ancient term used for a now unknown species of thorny shrub or tree.

Tobacco brush (left), *Ceanothus velutinus* Douglas ex Hook.

The tobacco brush has many other names, including mountain-balm, sticky laurel, buckbrush, greasewood, and just plain ceanothus. Recognize it by its foamy clusters of small flowers and leathery, three-veined leaves. The flowers have an odor reminiscent of tobacco, whence its common name. The plant is common in most of our western mountain states to subalpine elevations. It and the closely related *Ceanothus sanguineus* Pursh (redstem ceanothus, or Oregon teatree) are important in forest reclamation, for the plants spring up on burned ground where their seeds have remained dormant for years until activated by the heat of a wildfire.

Cascara (left and below), *Frangula purshiana* (DC.) Cooper

Cascara is a small tree with strongly ribbed deep green leaves, small white flowers and sparse blue fruit (berries). It is an attractive plant, and that is probably why Meriwether Lewis gathered it near today's Kamiah, Idaho, on May 29, 1806. Lewis made no mention of the bark's purgative value—apparently the region's Nez Perce Indians were unaware of this. Native Americans

in California knew about cascara, however, and told Spanish priests about its laxative effect—explaining why the bark is known medically as *cascara sagrada* (Spanish for "sacred bark"). The trees were scarce for a while in Idaho because so many were harvested for the bark. They are now grown commercially, and once again are found in fair numbers along the Clearwater River, where this one was photographed. The generic name, *Frangula,* was derived from the Latin *frangere* meaning "to break" for the brittle twigs of some species (cf. also our word "fracture").

173

Rose Family (Rosaceae)

The Rose family is made up of 110 genera and 3,100 species. It is well represented in our mountains by plants that range in size from tiny-flowered alpine plants (*Sibbaldia, Kelseya*) to full-sized trees. We have two species of wild rose (*Rosa* spp.) that anyone will recognize immediately; many other plants in the family, however, bear little resemblance to roses. There are, of course, similarities. Rosaceae flowers have in common five sepals united at the base of the flower to form a disk or cup (hypanthium or hypan) to which the petals (also usually five) and stamens (usually many) are attached; most species also have sepal-like leaves or bracteoles that arise from the stem just below the flowers. Fruit in the rose family takes many forms: drupes as in the genus *Prunus* (plums, apricots, almonds, peaches, cherries, etc.) and pomes as in *Malus, Pyrus, Cydonia* and *Eriobotrya* (apples, pears, quinces, loquats). Some members of the family form multiple small achenes (seeds) or drupelets, which mature into aggregate fruits as in *Rubus* (raspberries, blackberries and similar plants) and *Fragaria* (strawberries). Many of the Rosaceae are economically important as garden ornamentals. Roses, from which the family's scientific name was derived, have been known by that name, or a similar one in European languages as far back as ancient Greece (the island of Rhodes [*Rodos*] may have taken its name from a related, or similar, flower).

Shrubby cinquefoil
Dasiphora fruticosa (L.) Rydb.

Patches of shrubby cinquefoil are shown here growing far above treeline with Mount Borah's summit as a background. The plants are found in rocky depressions that offer a degree of protection from wind and cold. Shrubby cinquefoil adapts well to severe conditions and is at home in most of our northern states and throughout Canada. The plant blooms from early summer through August, so it is a favorite with landscape gardeners, who know it as "potentilla." Look for it on alpine tundra, in mountain meadows and in cities as an ornamental. *Dasiphora,* from the Greek, implies "thick foliage"; *fruticosa,* from the Latin, means "shrubby"; and the common name "cinquefoil," from the French, means "five-leaved" for the plant's five-fingered foliage. Meriwether Lewis collected a specimen of shrubby cinquefoil along Montana's Blackfoot River on July 6, 1806, on the expedition's return route.

Western serviceberry
Amelanchier alnifolia
(Nutt.) Nutt. ex M. Roemer

The western serviceberry (also known as Saskatoon serviceberry) is an attractive small tree that grows throughout northern North America to mid-elevations. The species name, *alnifolia*, means "alder-like leaf." Its deep blue berries are edible, although they have little taste. Lewis and Clark collected a variety of the western serviceberry, var. *semiintegrifolia*, at The Dalles in Oregon and then this plant, var. *alnifolia*, in north-central Idaho in the spring of 1806. The name *Amelanchier* is derived from a Savoyard term for the medlar-tree, *Mespilus germanica*.

Black hawthorn
Crataegus douglasii Lindl.

The black hawthorn is found from the Dakotas to the Northwest, growing to fairly high elevations. The tree is easily identified by its long, sharp thorns, rounded leaves with scalloped ends, clusters of white, spring-blooming flowers and later, by its dark red, drying-to-black fruit. The gnarled trees have a heavy bark and grow as much as thirty feet high. Palatable, but hardly delicious, "haws" were an important food for Native Americans. Both this plant, and the similar red hawthorn, *Crataegus chrysocarpa*, grow in Idaho, the latter as a cultivated plant.

Mountain ash
Sorbus scopulina Greene

Species of mountain ash—unrelated to the true ash of the eastern United States (*Fraxinus* spp.) are found in many parts of the West. This is a small tree that bears clusters of white flowers in late spring, ripening in late summer into colorful bunches of orange berries. Lewis and Clark collected a fruiting specimen on September 2, 1805, on the North Fork of the Salmon River during their journey west, and again on Lolo Pass on June 27, 1806, during their homeward journey. The closely related and very similar Sitka mountain ash, *Sorbus sitchensis*, is also found in Idaho, although at higher elevations.

Western chokecherry
Prunus virginiana L. var. *melanocarpa*
(A. Nelson) Sarg.

The chokecherry grows in all Canadian provinces and throughout the United States except for the Deep South. The western variety, var. *melanocarpa*—the term means "black fruit"—is a tall bush that flowers when it is only a foot or so tall. The plants usually mature as tall bushes; rarely, they grow as trees—we have seen them as much as twenty feet high, although this is unusual.

The eastern variety, var. *virginiana,* differs in that it is a sizeable tree that does not flower until it is mature. The tart but edible cherries are drupes, fleshy fruit with seed-containing pits. Lewis and Clark gathered specimens of western chokecherry twice, first in September 1804, in today's South Dakota, and again on May 29, 1806, while camped on the Clearwater River in today's Idaho.

American plum
Prunus americana Marshall

The American wild plum is widely distributed, found almost everywhere in the United States with the exception of Oregon, Nevada, Texas and California. The tree shown here was photographed in the foothills of the Clearwater Mountains near Harpster, Idaho.* The American wild plum is the tastiest of all our native *Prunus* species. As seen in the photographs, early blooming flowers are arranged in an attractive cluster that appears before the leaves are out. The fruit may be yellow, orange or even, occasionally, red. Its lanceolate, serrated leaves are typical of *Prunus* species in general.

---

*The trees in this region may have been planted by settlers; nevertheless, native American plums grow in western Montana, not far from where this specimen was photographed.

Bitterbrush
Purshia tridentata (Pursh) DC.

The bitterbrush (also known as antelope-brush) grows throughout the West. It can be confused with the shrubby cinquefoil, although it blooms earlier and has smaller, quite different flowers, and three-toothed (*tridentata*) leaves. The plant is usually found with sagebrush and may grow quite high in our mountains. It is an important browse plant for deer and antelope. The name *Purshia* honors Frederick Pursh (1774–1820), the botanist who classified specimens returned by the Lewis and Clark Expedition. Meriwether Lewis collected the plant on the same day as the shrubby cinquefoil.

Hillside ocean spray
Holodiscus discolor (Pursh) Maxim.

Each spring the Clearwater River Gorge is alive with ocean spray shrubs in bloom—and they would have been in bloom while Lewis and Clark were camped nearby, in today's Kamiah, Idaho. Most likely it was Meriwether Lewis who collected a specimen on May 29, 1806. Ocean spray grows throughout the West, from British Columbia to Texas and south to South America. As the illustration shows, it is an attractive spring-blooming bush whose dependent clusters of tiny off-white flowers make it a handsome ornamental plant.

Mallow-leaf ninebark
Physocarpus malvaceus (Greene) Kuntze

The mallow-leaf ninebark is a common shrub in our mountains, growing to fairly high elevations. Circumscribed clusters (corymbs) of flowers grow on the ends of many small branches; these, as well as the plant's rough and peeling bark (whence "ninebark") help to identify it. The species name, *malvaceus,* means "mallow-like,"as the leaves are similar to those of some mallows (although the leaves might also be referred to as "maple-leaf-like"). Interestingly, Meriwether Lewis several times mentioned seeing "nine-bark" (or "seven-bark"), a plant he knew from the East, undoubtedly the reason he did not collect a specimen.

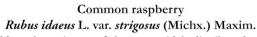

Common raspberry
Rubus idaeus L. var. *strigosus* (Michx.) Maxim.

The wild raspberry is one of the most widely distributed members of the rose family, found throughout North America (save in a few southern states) as well as in Eurasia. The plants are quite at home in our mountains and grow at least as high as treeline. Spring-blooming, five-petaled, small white flowers are typical of those of most *Rubus* species. The berries do tend to be smaller than domestic fruit—hardly surprising for they grow on unfertilized rocky ground. Nevertheless, if you sample the ripe fruit you will find that it is the same plant as the delicious raspberries that grow in our gardens.

Thimbleberry, *Rubus parviflorus* Nutt.

The thimbleberry is common in Idaho's mountains, growing as an "unarmed" (lacking brambles) shrub that may be six feet or more high. The plants are easily recognized by their large, deep green, maple-like leaves, by their large white blossoms (up to two inches across), and by their raspberry-like fruit. The fruit—unlike that of raspberries—is disappointingly tasteless. Lewis and Clark collected the thimbleberry on April 15, 1806, while near today's The Dalles, Oregon. Unfortunately, their specimen was in poor condition on its return to the United States, so it could not be published as a new species. Thomas Nuttall later found the plant on Mackinac Island, in Lake Huron, and gave it the species name *parviflorus* ("small flower"), a strange choice, for its flowers are the largest of any of our native *Rubus* species.

Wood's rose, *Rosa woodsii* Lindl.
var *ultramontana* (S. Wats.) Jepson

Wood's rose in full bloom is a most attractive plant. Its leaves and flowers are smaller than those of the Nootka rose (shown on the next page). While it grows high in the foothills, it is most commonly encountered as a lowland plant growing throughout the western United States and Canada, east to the Mississippi River. Several varieties are recognized. Another rose that is sometimes encountered in northern Idaho is the imported *Rosa canina,* the European dog-rose. The easiest way to distinguish it from native species is by the thorns. Those of the dog-rose are curved backward, whereas the thorns of the native species are straight.

Nootka rose
Rosa nutkana C. Presl.
var. *hispida* Fernald

The Nootka rose is named for Nootka Sound on Vancouver Island, where it was first collected by a botanist with the Malaspina Spanish expedition (1791), although the plant was not described until much later. Four varieties have been described; our plant's varietal name, *hispida*, means "bristly." It may grow to be six feet or more tall in well-watered places. The flowers of wild roses are quite similar, so other criteria are used to identify the various species. This plant's compound leaves have three to seven leaflets and are edged with fine teeth. The fruit, its "rose-hips," might serve as food in an emergency, but are seed-filled and unpalatable; they are better used as a vitamin C–rich tea.

Subalpine spirea
Spiraea splendens Baumann
ex K. Koch.

The subalpine spirea is a low shrub characterized by dense clusters of tiny pink flowers. Its deciduous leaves are simple and edged with fine teeth. The plant grows only in the mountains of the Northwest and in California and Nevada, usually on rocky ground and often in moist places. There are many species of spiraea; understandably, these attractive plants, this one especially, are often planted in ornamental gardens.

White spirea
Spiraea betulifolia Pallas

Less well known than its gaudy relative, the subalpine spirea, this plant has similar growth preferences. It is found on open terrain, usually close to streams, growing as high as the subalpine zone. The blooming shrub is attractive with its individual white flowers. Eastern and western varieties of *Spiraea betulifolia* have been described. Ours, var. *lucida* Dougl. ex Greene C. L. Hitchc., grows in northern states from Oregon and Washington to the adjacent Canadian provinces and east to Minnesota.

Prairie smoke
***Geum triflorum* Pursh**
var. *ciliatum* (Pursh) Fassett

Prairie smoke flowers, with their nodding, vase-like shape, reddish color and recurved bracteoles (accessory sepal-like leaves) are unique. The plants grow as high as treeline, blooming in mid- to late spring, often in large patches. *Geum* is an old Latin name for plants in this genus. The species name, *triflorum*, describes the plant's three-to-a-stem flowers. Imagine a patch of fruiting plants (right) and you'll see how the common names "prairie smoke" and "old man's whiskers" were derived. Lewis and Clark gathered this plant—previously unknown to science—near Idaho's Weippe Prairie on June 12, 1806. *Geum triflorum* grows all across northern North America as far east as Wisconsin and Ontario.

Ross's avens
***Geum rossii* (R. Br.) Ser.**
var. *turbinatum* (Rydb.)
C. L. Hitchc.

Ross's avens is a subalpine or alpine plant that is at home on rocky tundra. Its pinnatifid (feather-like) leaves help to identify the plant, as do its purple-tinged stems and calyces. Several varieties have been described, but this is the only one found in Idaho. The plant occurs in other western mountain states, also at high elevations. The species name, *rossii*, honors the arctic explorer Sir James Clark Ross (1800–1862). The varietal name, *turbinatum*, refers to the raised central disk.

Gordon's ivesia
Ivesia gordonii (Hook.) Torr. & A. Gray

Gordon's ivesia is found at high elevations throughout Idaho. The plant's pale yellow flowers are clustered into heads on the end of each of several long stems. The petals are glossy, giving the flowerheads an overall glistening appearance. The central portion of the flower—the hypan—is raised and often plumped up, or turbinate. The leaves are pinnate, made up of very closely ranked small leaflets. Once one knows the plant, it is surprising how often it is encountered growing on rocky alpine terrain. It is found in the northern and central Rocky Mountains, west to the coastal ranges. Four varieties occur in Idaho. The plant in the illustration, var. *ursinorum* (Jeps.) Ertter & Reveal, grows in the southern half of the state (the tall plant in the background is an unidentified composite). Dr. Eli Ives (1779–1861), for whom the genus was named, was a botanist and physician. A similar plant, *Ivesia tweedyi* Rydb., is rare; it grows in mountains near Coeur d'Alene, Idaho.

Kelseya
Kelseya uniflora (S. Watson) Rydb.

The kelseya is an early flowering, mat-forming plant that grows above treeline in the mountains of Idaho, Wyoming, Colorado and Montana (where it also grows at lower altitudes). It is the only plant in its genus. Surprisingly, even though the kelseya is uncommon, it has been cultivated as an "alpine" in rock gardens throughout the world. One can understand why it is popular with rock gardeners, for it is an exotic and attractive little plant that spreads over rocks, forming a carpet of tiny blue-gray leaves dotted with minute pink and white flowers. The flowers—shown greatly magnified here—seem to be no larger than the head of a match.

Kelseya uniflora was named for Rev. Frank Duncan Kelsey (1849–1905), a resident of Helena, Montana (and later of Toledo, Ohio), who first collected the plant.

Cliff drymocallis
Drymocallis pseudorupestris (Rydb.) Rydb. var. *saxicola* Ertter

Until recently this wildflower was classified as *Potentilla glandulosa* Lindl. with the common name "sticky cinquefoil." Recent studies, however, have shown that it is not related to the potentillas—not a surprising finding given its white flowers and pinnate leaves. Var. *saxicola*, common in Idaho, grows as high as its treeline, where its size is reduced (left). The species, as one or another of many varieties, is found in most states and provinces west of the Mississippi River. The name *drymocallis*, from the Greek, means "wood beauty"; the Latin varietal name, *saxicola*, means "mountain (or cliff) dwelling," reflecting this variety's growth preference.

Early cinquefoil
Potentilla concinna Richards.

The early cinquefoil is one of several quite similar, low-growing potentillas that are found high in our mountains. Its leaves vary from digitate (as shown here) to pinnate. The individual leaflets are toothed at the ends. It may be distinguished by its leaves, for their undersurfaces are covered with downy, whitish hairs—they are tomentose. The early cinquefoil grows from the central Canadian provinces, through the Rocky Mountain states to Arizona and New Mexico, west to Nevada and east to the Dakotas. It was collected in 1820 by John Richardson, physician-naturalist with the first Franklin expedition, at Fort Carlton on the Saskatchewan River in Canada. The name *concinna*, from the Latin, means "neat."

Mountain meadow cinquefoil
Potentilla diversifolia Lehm.

As the scientific name suggests, there is considerable variation in this plant's leaves and its leaflets. Although our photograph does not show this well, it takes the form chiefly in the degree to which the divisions between the leaflets are split or dissected. Some leaves show little differentiation, others are strongly divided into five or more digitations (think of an analogy with fingers given off a palm). The plant in its several varieties is widespread and may be found in most of the western states and north to Alaska. Separate populations have also been found in Labrador and Greenland.

Slender cinquefoil
Potentilla gracilis **Douglas ex Hook.**
var. *brunescens* **[Rydb.] C. L. Hitchc.**

The slender cinquefoil is one of the most variable of all potentillas. It grows throughout the West, as far east as the Dakotas, all across Canada and into Alaska. There are several varieties (var. *brunescens* [Rydb.] C. L. Hitchc. is shown here) characterized by differences in their leaves—how deeply indented the lobes of the leaflets are, how much hair grows on their surfaces, etc.—differences mostly of importance to botanists. All have palmately compound leaves with five to nine toothed, pinnate (feather-like) leaflets and five-petaled yellow flowers with a central disk, typical of the rose family. The species name, *gracilis*, means "slender."

Biennial cinquefoil
Potentilla biennis **Greene**

The biennial potentilla takes the form of a rather skinny, single-stemmed plant with distinctive, smoothly serrated leaflets that grow closer together as they approach the top of the stem where a cluster of yellow flowers, typical of potentillas in general, appears. The plant is listed as not being present in Idaho, which would be surprising if true, for the biennial cinquefoil is found in every one of the states and provinces surrounding Idaho. (The plant shown here was photographed on Sun Valley's Dollar Mountain.)

Pinewoods horkelia
Horkelia fusca **Lindl. var.** *parviflora*
(Nutt. ex Torr. & Gray) Keck

Until recently *Horkelia parviflora* was classified as its own species; now, however, it is considered to be a variety of *Horkelia fusca*. The generic name honors German plant physiologist and physician Johann Horkel (1769–1846); the species name, *fusca*, is from the Latin and means "dark-colored" or "brown." It is a meadow plant that grows to fairly high elevations in our mountains, usually forming patches of brown-stemmed plants set off by clusters of small white flowers. The plant is found in the four northwestern states, California and Nevada.

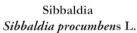

Sheep cinquefoil
Potentilla ovina
Macoun ex J. M. Macoun

The sheep cinquefoil has—at least for the amateur botanist—a good thing going for it; its leaves are unusual for a potentilla, making it easy to recognize. It is a small creeping plant that grows from montane to alpine elevations. Pinnate leaves arise mostly from the base of the plant and have small, tightly ranked, furred leaflets. The yellow flowers are similar to those of other potentillas. It is said that Macoun, who described this plant, found it on Sheep Mountain in British Columbia and gave it the name *ovina*, a word that means "of sheep." The plant is found all through the Rocky Mountains and west to mountain ranges in Oregon.

Sibbaldia
Sibbaldia procumbens L.

Sibbaldia procumbens is the only North American plant in its genus (although five other species occur in Eurasia). As is often the case, when the author of one of our plants is designated as "L." (for Linnaeus), the plant is found in both the Old and New Worlds. The sibbaldia is an alpine plant in the United States, growing at progressively lower elevations north to Siberia and Alaska, across Canada to Labrador and Greenland. Sibbaldias (the common name "creeping-glow-wort" has also been suggested) are dwarf plants, far smaller than the illustration suggests. They may be identified by tiny flowers borne in small clusters along and at the ends of stout stems. The flowers have five prominent sepals and five narrowly attached petals. Basal leaves are strawberry-like, with several teeth at the end of each of three leaflets. The plants spread by creeping rhizomes. Sir Robert Sibbald (1641–1722) was a prominent physician of Edinburgh with a strong interest in botany. He published a natural history of Scotland in which he described various plants, including this one. When Linnaeus described it, he took note of Sibbald's contribution, naming it *Sibbaldia*.

Wild strawberry
Fragaria virginiana Mill var. *glauca* (S. Watson) Staudt

Wild strawberries are found in every part of North America. Common in our mountains, they occur at all elevations, even above treeline. Their flowers are large, with five separated petals and many anthers and stigmas arising from a central receptacle. Fruiting plants are not common, but when encountered are unmistakably strawberries, both in taste and appearance. Two varieties occur in Idaho. The plant shown here has smooth, almost hairless leaves. The other variety, var. *platypetala*, has larger flowers and furry leaves and stems. The generic name, *Fragaria*, was derived from *fraga*, the Latin word for "strawberry."

Salad burnet, *Sanguisorba minor* Scop.

Although several native species of burnet are found in Idaho, this one is an introduced, supposedly edible plant. It is characterized by twelve pale anthers protruding from flowers at the bottom of the flowerhead; the female flowers with reddish stigmas are higher up within the flowerhead. The plant has fern-like, compound, pinnate leaves (not present in our image). The burnet grows in open fields as high as the montane zone. The common name "burnet" has been used for centuries; it means "brown," the color of post-mature flowerheads. The name *Sanguisorba* implies "blood absorbing," for in the past the plants were thought to have styptic properties. Our burnet is found everywhere except in the states of the Great Plains and the Deep South.

Madder Family (Rubiaceae)

The Rubiaceae family is moderately large--approximately 650 genera and some 13,000 species; only a few species are found in Idaho. Most—trees, shrubs, vines and a few herbaceous plants—are native to the tropics. Family characteristics include square stems, opposing narrow leaves usually arranged as whorls at intervals along the stem, and small four- or five-petaled flowers. Some family members contain alkaloidal substances that are important to man. These include quinine (*Cinchona* spp.) and ipecac (*Psychotria emetica*)—both from South American trees; yohimbine derived from an African tree (*Corynanthe johimbe*), used to treat impotence and as a typical vasodilator in surgery; and coffee from the fruit of various species of *Coffea* (Africa). Some of our most important garden ornamentals are also in the Madder family: *Gardenia, Penta, Ixora* and others. Finally, the red dye madder, from which the family takes its common name—and indirectly its scientific name—is also obtained from species of Rubiaceae.

The word "madder" has been applied to a red vegetable dye throughout history. While the dye is found in the roots of other Rubiaceae (*Galium, Asperula*), the greatest concentrations occur in species of *Rubia*, from which the family name Rubiaceae was derived (*rubus* is Latin for "red"). Dyer's madder, used to dye fabrics, and the artists' pigment "rose-madder" was extracted from the European plant *Rubia tinctorum*, and from *Rubia cordifolia*, an Asian plant. In the second half of the nineteenth century, madder dyes—alizarin and purpurin—were synthesized, so plant madder is seldom used today. The bedstraws (*Galium* spp.) are sweet-smelling Rubiaceae that dry to give a stuffing used for pillows and mattresses, whence their common name. "Cleavers" is another name for bedstraws, one that conjures up mental pictures of a wicked kitchen utensil. "To cleave" is an Old English word, however, it means "to stick" or "adhere" (cf. the injunction "cleave unto me"), because the seeds, stems and branches of some bedstraws attach themselves to animal and human passersby.

Northern bedstraw, *Galium boreale* L.

The northern bedstraw is a ubiquitous circumpolar plant found everywhere in the United States (except the Deep South, Kansas, Oklahoma and Arkansas), in almost all the Canadian provinces and Alaska, and in Greenland and Eurasia. The plants are many-branched. Four lanceolate leaves arise from nodes along the stems. Dense clusters (cymose panicles) of small, four-petaled, white flowers are borne on stemlets given off from leaf nodes at the top of the stems, producing an overall showy profusion. The plants grow at all elevations, from sea level to treeline, preferring moist situations. The foliage is said to be edible. (The background yellow composite is a species of *Arnica*.)

Fragrant bedstraw
Galium triflorum Michx.

The fragrant bedstraw rambles along forest floors, loosely climbing other plants that it encounters. Its bright green, narrow leaves form whorls at intervals along the stems. Small four-petaled flowers, inconsistently borne in groups of three at the end of long stemlets, arise at the leaf nodes. They give off a sweetish grassy odor; one can understand how the dried plants of this and related species might make pleasant-smelling bedstraw for pillows and mattresses. The plant's appearance is similar to that of the related European herb, sweet woodruff or waldmeister (*Galium odoratum* Scop.), that is steeped in white wine to make May wine. *Galium triflorum* is a circumboreal plant that is found throughout North America and as far south as Mexico.

Watson's bedstraw
Galium watsonii (A. Gray) A. Heller

Watson's bedstraw is encountered during the summer in high, dry places. The four-petaled flowers are small and unremarkable save for the bristly hairs that grow on the flowers' ovaries. These give the female flowers a brush-like appearance (the plants are dioecious—the male plants are not so hairy). The name *Galium* was derived from the Greek *gala* for "milk" (cf. "galaxy" for the milky way), because the European yellow bedstraw (*Galium verum* L.) curdles milk and formerly was used as a vegetable rennet in cheese-making. The plant shown here is found in the three Pacific coastal states, east to northwestern Nevada and into Idaho.

Intermountain bedstraw
Galium serpenticum
Dempster & Ehrend

Galium serpenticum is a semi-woody shrub made up of stems, leaves and flowers so tangled up as to remind one of a den of hibernating snakes (left, in the image above). It is more likely that its species name refers to the serpentine rock formations on which the plant often grows. The species has been found growing in the three Pacific coastal states and in Nevada and Idaho. Nine varieties have been described. In order to show the plant's morphology better, we used a photo-editing program to isolate a single stem (right, in the image above). The whorl of leaves is typical of *Galium* plants, as are the small four-petaled flowers.

Saxifrage Family (Saxifragaceae)

The word "saxifrage" is derived from the Latin *frango* meaning "I break" and *saxum* meaning "rock." Pliny (born AD 23, died AD 79 in the eruption of Mount Vesuvius) in his *Natural History*, wrote that the name *saxifragum* ("stonebreaker") referred to the plant's supposed ability to dissolve kidney and bladder stones. More likely, however, it reflected the growth habit of many of the plants in the family, for saxifrages often grow in rocky clefts, seemingly having broken the stone apart. The family is not a large one. It consists of about thirty genera and 325 species. The plants are mostly herbaceous (non-woody). Although the family is distributed throughout the world, most members originate in the north temperate zone, often in inhospitable surroundings (desert, arctic, alpine, bogs and as aquatic forms). The family has little economic importance, although a few are used in folk medicine and others find a place in ornamental gardens. Saxifrage flowers usually have five sepals, five petals and ten stamens. As will be seen on the following pages, many plants in this family have, as a useful identifying feature, clusters of small flowers borne atop a long stem with a basal rosette of toothed leaves.

Diamond-leaf saxifrage
Saxifraga rhomboidea Greene

This saxifrage is a neatly symmetrical, usually stand-alone plant. The bare stem emerges from a basal rosette of more-or-less diamond-shaped, serrated leaves. Its flowerhead is compact and usually somewhat globose. The stout stem is smooth to slightly hairy. Diamond-leaf saxifrages prefer moist, rocky ground where they bloom from late spring into the summer, depending on the elevation. It is found from the Canadian province of Alberta through the Rocky Mountain states as far south as the mountains of Arizona.

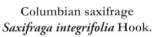

Columbian saxifrage
Saxifraga integrifolia Hook.

Although the U.S. Department of Agriculture Plants Database (http://plants.usda.gov) lists this plant as not being found in Idaho, it is present at least in central Idaho, and we suspect elsewhere in the state as well. The Columbian saxifrage is also found in the three coastal states as well as in British Columbia and Montana. It is similar to many other plants in genus *Saxifraga* except it is lacking the toothed or serrate leaves that are present in most plants in this genus. That also explains the species name, *integrifolia*, meaning entire leaf, and is a helpful feature in identifying the plant.

Brook saxifrage
Saxifraga odontoloma Piper

The brook, or streambank, saxifrage, formerly *Saxifraga arguta*, is also typical of plants in the genus *Saxifraga* with its basal cluster of leaves and long stem bearing many small flowers. The brook saxifrage is a common plant in our mountains, turning streambanks green. The plants are easily identified by their attractive, bright green, deeply scalloped, oval or fan-shaped leaves. These give the plant both its old and new species name (both mean "toothed"). The brook saxifrage is able to thrive in varying light intensities, growing equally well in bright sunlight or deep shade. It is found in all the western mountains from Alaska southward to Mexico.

Mountain saxifrage
Saxifraga occidentalis S. Watson

In the past, this species was made up of a half-dozen varieties, each with slightly different characteristics and territories. Recently, however, these have been reclassified, leaving *Saxifraga occidentalis* as its own, well-defined, species. The plant blooms in early summer, preferring the moist ground of mountain meadows and the banks of mountain lakes. It may be identified by a stem that gives off many stemlets, each bearing a cluster of white flowers. Each flower has two prominent red carpals that mature into a two-parted capsule (follicle). The plant is native to the Rocky Mountains, and south to New Mexico, as well as in the four northwestern states. It is also found in Alberta and British Columbia.

Purple saxifrage
Saxifraga oppositifolia L.

This lovely plant is one that few people see, for it is a circumboreal plant that grows only on alpine tundra and is found in only a few places in Idaho. Two small, opposing leaves (*oppositifolia*) are tightly arrayed to form a leafy pillar that lengthens a bit each year. Striking pink to purple flowers up to an inch in diameter bloom at the end of these. The plants shown here were growing high on Mount Borah in Idaho's Lost River Range. Farther north, in the arctic of both New and Old Worlds, purple saxifrages may be seen growing on arctic tundra at elevations as low as as sea level.

Fringed grass-of-Parnassus
Parnassia fimbriata K. D. Koenig

Parnassia species are circumboreal, subalpine and alpine plants that grow at high elevations throughout the Northern Hemisphere. *Parnassia palustris* is the original grass-of-Parnassus found both in Eurasia and North America (including Idaho). *Parnassia fimbriata,* shown here, is identical except for its petals whose edges are fringed (*fimbriata*) for half of their length. It blooms, from midsummer on, in swampy mountain meadows and along the banks of slow-moving mountain streams at montane to alpine elevations. It is easily identified, for the plants have single white flowers atop naked stems that emerge from basal clumps of kidney-shaped leaves.

Gooseberry-leaved alumroot
Heuchera grossulariifolia Rydb.

The gooseberry-leaved alumroot blooms from late spring well into the summer, favoring cliffsides (where it usually grows as a solitary plant) and rocky ground (where it may grow in clusters) from mid-elevations to alpine tundra. The species name is derived from the resemblance of its leaves to those of gooseberries in the currant family (Grossulariaceae).* Its small flowers have five sepals joined to form a bell-shaped receptacle that almost hides the petals. The popular name "alumroot" is derived from its puckery taste, for the roots and stems contain a high concentration of tannin. The generic name honors Johann Heinrich von Heucher (1677–1747), professor of medicine at Wittenberg in Germany. It is unlikely that Heucher knew this plant, for it is not found in Europe. Linnaeus, in his *Species Plantarum* (1753), often gave plants the names of prominent men who had nothing to do with their namesakes. This, and the poker alumroot shown below, are both plants of the northern Rockies and Pacific coastal ranges.

———————

*At one time, members of the Gooseberry family, Grossulariaceae, were classified as Saxifragaceae, so it is not surprising that the plants bear some resemblance.

Poker alumroot
Heuchera cylindrica Douglas ex Hook.

The poker alumroot is quite similar in appearance and in growth habit to the gooseberry-leaved alumroot shown above, differing mainly in the shape of its leaves. These tend to be round with shallow lobes. There is considerable difference in this plant's morphology from place to place and even from plant to plant, and six different varieties are recognized; these have relatively minor variations in leaf shape. Unlike *Heuchera grossulariifolia* shown above, petals are sometimes absent in the flowers of this species.

191

Smallflower woodlandstar
Lithophragma parviflora (Hook.) Nutt. ex Torr. & A. Gray

The woodlandstars (also known as prairiestars) illustrated here are two of the three species often seen in our mountains. All are quite similar. The showy, white (occasionally pale pink) flowers are only about half an inch in diameter, but they are eye-catchers, thanks to their cleft petals. Prominent calyces with five pointed tips cup the flowers. Five deeply cleft petals have, depending on the species, three, five or seven lobes. Small flower clusters are borne atop long, usually reddish- to purple-colored stems that arise from a basal cluster of lobed leaves. Woodland stars are montane plants, scattered in generous numbers in sagebrush and mountain meadows, and along streambanks, where they bloom into the summer at higher elevations. This species is found in British Columbia and Alberta, south to California, Nevada, central Utah, northern Colorado and northwestern Nebraska.

Bulbous woodlandstar
Lithophragma glabrum Nutt.

The bulbous woodlandstar (also known as prairiestar) was until recently classified as *Lithophragma bulbifera* Rydberg. The plant often forms tiny bulbs at the base of the flowers and leaves, and these sometimes replace those structures. The bulbous prairiestar's petals are rather attenuated and usually have five deeply dissected lobes, unlike the woodlandstar shown above. It also tends to bloom earlier in the spring than other species, and it is not unusual for the petals to have a pinkish tinge. The genus name, *Lithophragma*, was derived from the Greek *lithos* for "stone" and *phragma* meaning "wall," from the plants' tendency to grow in rocky places. While bulbils are common in this species, other woodlandstars may, on occasion, also produce them. In addition to those that form aboveground, subterranean bulbils may also form, and are probably more effective in creating new plants than are aboveground bulbils and seeds.

Five-stamen mitrewort
Mitella pentandra Hook. (left, top)
Side-flowered mitrewort
Mitella stauropetala Piper (left, center)

The five-stamen (or alpine) mitella, *Mitella pentandra* and the side-flowered mitella, *Mitella stauropetala* (from *stavros,* a Greek word for "cross," reflecting the shape of the attenuated petals) are two of the four species of *Mitella* found in Idaho. The side-flowered mitrewort grows in Idaho, west to Oregon and Washington, east into Montana and Wyoming, and south into Utah and western Colorado. The five-stamen mitella has much the same distribution in the United States but also ranges as far north as Alaska and the neighboring Canadian provinces, and south to California and Nevada. Both prefer moist environments and are usually found in shaded woods and along streambanks. As with many other plants in the saxifrage family, the plants' basal leaves are disproportionately large when compared to the tiny flowers (the flowers are greatly magnified in these illustrations). The bizarre little flowers with their five skeletal petals should be easy to identify the first time they are seen. The word "mitella" is a diminutive form of the Latin *mitra* ("hat" or "cap"), a reflection of the shape of the plants' fruiting body said to resemble a bishop's hat.

Threeleaf foamflower
Tiarella trifolia L. var. *unifoliata* (Hook.) Kurtz

The threeleaf foamflower (also known as the coolwort and the laceflower) ranges from Alaska, south to California, east to Alberta and across Idaho to Montana. It too prefers deep woods and well-shaded streambanks. Tiny, white, five-petaled flowers are borne in small clusters (panicles) on stemlets arising from a single long stem. The varietal name, *unifoliata,* differentiates this variety from the very similar var. *trifoliata,* in which the leaves are divided into three separate leaflets, rather than the single, lobed leaves of the plant shown here.

Figwort, or Snapdragon, Family (Scrophulariaceae)

Because the figwort, a Eurasian plant, is not well known in America, the Scrophulariaceae family is sometimes referred to as the Snapdragon family. Recently, with the advent of DNA studies, many plants previously classified as Scrophulariaceae have been reassigned to other plant families. Understandably, this is confusing for those who grew up using the older classifications. Despite these changes, various plants formerly classified as Scrophulariaceae do have common characteristics. These include flowers that have five petals (a few have four) joined for much of their length, a "lip" formed by prominent lower petals, and usually four anthers (although some genera—notably *Penstemon*—have a sterile fifth stamen).

The family, as originally classified, is a large one that included approximately 5,100 species spread throughout the world. Despite its size, its members have little economic importance other than as ornamentals. *Digitalis purpurea* is an exception; not only is it a popular garden plant, but the leaves are the source of digitalis, a potent and useful cardiac stimulant.

In order to simplify taxonomic matters as much as possible, we have elected to place all the plants previously classified as Scrophulariaceae under that heading, while including the newer—now mostly accepted—classifications as subheadings. Of the plants shown here, these include:

Orobanchaceae (Broomrape family) *Orobanche, Castilleja, Orthocarpus, Pedicularis* and *Cordylanthus*

Phrymaceae (Lop-seed family) *Mimulus*

Plantaginaceae (Plantain family) *Penstemon, Collinsia, Veronica, Synthyris, Chionophila* and *Linaria*

Scrophulariaceae (Snapdragon family) *Verbascum* and *Limosella*

Western naked broomrape, *Orobanche uniflora* L. var. *occidentalis* R. L. Taylor & MacBryde

The naked broomrape is a non-chlorophyllaceous plant found throughout North America. It is a parasite whose modified root system invades the roots of neighboring plants—usually sagebrush. Lack of chlorophyll results in a yellowish-brown to purple coloration (the latter, as in the inset, is more noticeable at subalpine elevations). The plant's relationship to other hemiparasitic Scrophulariaceae (*Castilleja, Orthocarpus, Pedicularis,* etc.) has long been noted. These genera have been reclassified and are now in the Broomrape family. The "rape" in "broomrape" was derived from a Latin word, *rapum,* meaning "knob," referring to lumps that form on the roots of brooms (shrubs in the pea family), caused by a European broomrape. *Orobanche,* in turn, was derived from two Greek words and means, approximately, "vetch-strangler."

Scarlet paintbrush
Castilleja miniata Douglas ex Hook.

Most Indian paintbrushes are found in the American West; about forty species grow in the Northwest, too many to show more than a representative sampling here. Look closely at the lower photograph on this page and you will see the plant's terminal red bracts (specialized upper leaves). Among them are slim yellow flowers. Each has a four-spiked calyx from which five petals protrude. Two lower petals join to form a characteristic "lip"; there are also two tiny lateral petals and an overhanging beak-like fifth petal known as a galea (Latin for "helmet"). Paintbrushes are hemi-parasites. They are capable of photosynthesis, but also require root contact with another plant, usually sagebrush in our area. Lacking that, they are difficult or impossible to grow in cultivation.

Castilleja miniata, shown here, is the most common of Idaho's Indian paintbrushes. It grows at all elevations and thrives well above treeline on alpine tundra. The red bracts each have two small lateral projections—an identifying feature for this plant. David Douglas first collected the scarlet paintbrush in Oregon's Blue Mountains. The genus name, *Castilleja,* honors Spanish botanist Domingo Castillejo (1744–1793).

Other wildflowers in the upper illustration include mountain bluebell (*Mertensia ciliata*), Lewis's monkeyflower (*Mimulus lewisii*), a species of *Arnica,* monkshood (*Aconitum columbianum*), upland larkspur (*Delphinium nuttallianum*), brook saxifrage (*Saxifraga odontoloma*) and mountain death camas (*Anticlea elegans*). While not all are well shown here, all are described and illustrated elsewhere in this book. The trees are Douglas fir (*Pseudotsuga menziesii*) and subalpine fir (*Abies lasiocarpa*).

Rosy Indian paintbrush
Castilleja rhexifolia Rydb.

The rosy, or rhexia-leaved, Indian paintbrush may be identified by its dark red bracts and by its lanceolate leaves that lack projecting lobes. While not uncommon, the rosy paintbrush is seen less often than the scarlet paintbrush shown on the previous page. The species name, *rhexifolia,* links the shape of its leaves to those of rhexias, plants found mostly in the southern and eastern states. The rosy paintbrush grows from British Columbia and Alberta to northeastern Oregon, then south in the Rocky Mountain states to New Mexico. It is usually found in a fairly moist environment.

Wavyleaf Indian paintbrush
Castilleja applegatei Fernald

The wavy leaves (not shown well in the image above) of wavyleaf Applegate's Indian paintbrush are a distinguishing feature of this plant. It is widespread throughout the West, ranging from Montana, Wyoming, Colorado and New Mexico to the east, and west to the coastal states (except Washington). Given its widespread distribution, it is not surprising that half a dozen varieties have been recognized. The one pictured above appears to be var. *pinetorum,* from a Latin word meaning "of pine forests." The name is not uncommonly used for other plants, including several others in this book.

Rocky Mountain paintbrush
Castilleja covilleana L. F. Hend.

The Rocky Mountain, or Coville's, paintbrush is a localized plant found only in central Idaho and, unusually, in adjacent Montana. Despite its limited range, it is not an uncommon plant in the central part of the state. The plant is not hard to identify given its elongated, spidery, three-lobed leaves and bright red, orange or occasionally yellow bracts. It is one of our earliest blooming paintbrushes, at home on rocky, sagebrush-covered slopes. The plant's species name honors a prominent botanist, Frederick Vernon Coville (1867–1937), curator of the U.S. National Herbarium and chief botanist of the USDA.

Lemon-yellow Indian paintbrush
Castilleja flava S. Watson

Yellow is not an uncommon color for Indian paintbrushes, either as a primary color, as with this plant, or as an alternate (and confusing) color for others. The Latin word *flava* means "yellow" as do the words *fulva* (signifying a tawny yellow) and *sulphurea;* doubtless there are others as well. One needs to look for other characteristics for help in identifying yellow-colored paintbrushes. Clustered and sometimes branching stems, bracts with two long, narrow accessory lobes, and a long flower calyx also help with the identification of this plant. Unfortunately, a tendency to hybridize with other yellow paintbrushes can cause confusion. *Castilleja flava* is a common plant in the inter-mountain states; it also grows as far west as Oregon, where it is uncommon.

Cusick's Indian paintbrush
Castilleja cusickii Greene.

Cusick's paintbrush is a common montane to alpine plant in Idaho, the surrounding states and British Columbia. The plant's attractive bright yellow bracts, its preference for the mucky ground of moist meadows and its tendency to grow in discrete clusters are usually enough to identify the plant on first encounter.

Narrow leaf (or Northwestern) Indian paintbrush
Castilleja angustifolia (Nutt.) G. Don

The narrow leaf Indian paintbrush is an eyecatcher because of the striking and unusual pink color of its bracts—and below them, its leaves. The plants may grow to a considerable size as subshrubs. It is rather restricted in its location, found in Idaho and all the contiguous states save Washington and British Columbia (although it does occur in Alberta).

Thin-leaf owl-clover
Orthocarpus tenuifolius (Pursh) Benth.

Owl-clovers (also known as owl's clover) are closely related to Indian paintbrushes, so much so that at one time they were included in the same genus. Most owl-clovers are not particularly attractive, but this one with its delicate yellow and light purple bracts is an exception. It is a montane to alpine plant found in the northern part of Idaho as well as in other northwestern states and British Columbia. Meriwether Lewis collected this plant on July 1, 1806, while camped at "Traveler's Rest" near today's Missoula, Montana. *Orthocarpus* means "straight fruit," referring to the shape of the seed capsule; *tenuifolia* means "thin leaf"—the plant's common name. The origin of the term "owl-clover" is obscure.

Elephant-head
Pedicularis groenlandica Retz.

The elephant-head lousewort (also known as pink elephants and Greenland lousewort) grows in all our western states and throughout Canada (although it is not found in Greenland). Its bizarre little flowers are borne in a tightly packed spike above feather-like leaves. The flowers have a hooded upper petal, or galea with a long projection that resembles an elephant's head and trunk. A three-lobed lower lip forms the face. It is a montane to alpine plant that grows on moist ground from late spring to midsummer. Louseworts, like other Orobanchaceae, are hemi-parasites. Meriwether Lewis collected the Greenland lousewort on July 6, 1806, on the Blackfoot River, upstream from today's Missoula, Montana.

White sickletop lousewort
Pedicularis racemosa Douglas ex Benth.
var. _alba_ (Pennell) Cronquist

The sickletop lousewort (also known as parrot's beak lousewort) is another *Pedicularis* species commonly encountered in our high mountain meadows. It grows in most of our western mountain states and adjacent Canadian provinces. The plant shown here, var. *alba*, is white-flowered and common to Idaho, although a pink to purple variety, var. *racemosa*, grows farther to the west. One will have no problem identifying either variety, for the galea (a modified upper petal) is sharply hooked into the sickle shape, as shown in the illustration, a shape that is responsible for its common names. The species name, *racemosa*, is a description of how the flowers are clustered, botanically as a "raceme."

Bracted lousewort
***Pedicularis bracteosa* Benth. var. *siifolia* (Rydb.) Cronquist**

The yellow-flowered bracted lousewort (left and foreground above) grows in moist places, at higher altitudes. The species name, *bracteosa,* is derived from small toothed leaves (bracts) that subtend each of its flowers. The plants have fern-like leaves, common to the genus *Pedicularis*. Seven varieties of bracted lousewort are recognized. The plants, as one variety or another, occur in the two western Canadian provinces, south to New Mexico and west to the mountains of Oregon, Washington and California, where it is uncommon.

Above: The bracted lousewort is shown with a medley of alpine plants (not all easy to make out). These include shooting stars (*Dodecatheon jeffreyi*), rosy paintbrush (*Castilleja rhexifolia*), pink mountain heather (*Phyllodoce empetriformis*), bistort (*Bistorta bistortoides*) and elephant-head lousewort (*Pedicularis groenlandica*)—all described in this book.

Yakima bird's beak
Cordylanthus capitatus
Nutt. ex Benth.

The two bird's beaks shown here, while not rare, are not particularly common plants that grow in our mountains to montane elevations or higher. The bird's beaks are not well known, presumably because they are less common and not nearly as striking as the closely related Indian paintbrushes. Examine the bird's beaks closely and you will see the lobed leaves and narrow, bracted flowers that are characteristic of various other plants in the Broomrape family (Orobanchaceae). Look for this one on sage-covered slopes in our northwestern states, where they appear in midsummer.

Bushy bird's beak
Cordylanthus ramosus
Nutt. ex Benth.

The bushy bird's beak with its yellow-brown, narrow-leaved foliage and flowers is even less conspicuous than is the Yakima bird's beak shown to the left. Although about eighteen species of *Cordylanthus* are found in our West, only the two shown here grow in Idaho. This one grows throughout the Northwest, south to California, Nevada, Utah, Wyoming and Colorado. Look closely and you will see its compact, elongated flowers and three-lobed leaves also found in many of the Indian paintbrushes. The generic name, *Cordylanthus,* was derived from two Greek words meaning "club flower" for the shape of the clustered bracts surrounding the narrow and inconspicuous flowers.

Owyhee mudwort
Limosella acaulis **Sessé & Moc.**

If one were asked to choose the most insignificant wildflower in this book, the Owyhee mudwort would certainly be a contender. The flowers are less than an eighth of an inch in diameter and obscurely bilaterally regular with two slightly larger petals. The flowers, their hairy buds and stemless (*acaulis*) leaves appear, often in vast numbers, on muddy places seemingly overnight, disappearing as the ground dries. Originally thought to be found only in south-western Idaho (Owyhee country!), it is now known to be far more widespread, occurring also in California and elsewhere in the Southwest. We suspect that it will turn out to be even more widely distributed than that. So far as we can determine, limosellas remain in the Figwort family.

Lewis's monkey-flower
Mimulus lewisii Pursh

Lewis's (also known as purple) monkey-flower has large, bright pink to almost purple blooms accentuated by a red-spotted yellow palate on the central petal of the lower lip. Its species name honors Captain Meriwether Lewis, who was the first to collect the plant on or near Lemhi Pass on the border of today's Montana and Idaho, in August 1805 (it still grows there today). It is also found in California and in the two westernmost Canadian provinces. The plant grows along mountain streams and in seep-springs, blooming at least as high as treeline. The flowers are shed whole and float into backwaters, often turning the surface pink with floating blooms.

Subalpine monkey-flower
Mimulus tilingii Regel

The subalpine monkey-flower spreads by shallow roots (rhizomes) and grows near water at high elevations. Its flowers have a lower "lip" made up of conjoined petals and a furry, variably red-spotted "palate." The plant grows from New Mexico, west to the coastal ranges of the United States and Canada. Heinrich Sylvester Theodor Tiling (1818–1871), a physician-botanist with the Russian American company, collected the plant's seeds near Nevada City, California, in 1868. The flowers of the common yellow mimulus, *Mimulus guttatus* DC. (not shown), look so much like those of this plant that the two species are hard to tell apart (although genetically separate). The yellow mimulus is a taller plant that grows at lower elevations. It thrives where there is moisture, and has been introduced far from its native West, throughout North America and elsewhere as well. Meriwether Lewis found the yellow mimulus near present-day Missoula, Montana, on July 4, 1806.

Dwarf purple mimulus, *Mimulus nanus* Hook. & Arn.
The dwarf monkey-flower is a foothill plant found in the Northwest and in Nevada and California. Its striking little flowers have variable and distinctive dark markings. The plants may grow in such numbers as to turn the ground purple.

Primrose monkey-flower, *Mimulus primuloides* Benth.
The flowers of the primrose monkey-flower are less than a quarter of an inch in diameter with crimson-spotted, notched petals. This montane to subalpine plant spreads by rhizomes (shallow roots) to form spreading, dense mats on wet ground. It is native to all the far western states.

Musk-flower, *Mimulus moschatus* Douglas ex Lindl.
The musk-flower, named for its odor, is a plant of the Northwest (although, interestingly, a disjunct population grows on the northeastern seaboard, north to Labrador). Four varieties are recognized, although var. *moschatus* shown here is the only one native to Idaho's mountains. The plants usually grow near streams or in other moist places. Because the flowers are obscurely lipped and the petals have no red markings, the plants may not be recognized as monkey-flowers when first seen.

Suksdorf's monkey-flower, *Mimulus suksdorfii* Gray
The tiny red-spotted flowers of Suksdorf's mimulus, greatly magnified here, measure only about an eighth of an inch across. These, as well as its reddish calyces, help to identify the plant. The one in the illustration below was photographed at the Craters of the Moon National Monument, where it is commonly encountered. Suksdorf's monkey-flowers are native to most states west of the Dakotas and Texas.

Blue-eyed Mary
Collinsia parviflora Douglas ex Lindl. (above, left and right)

Blue-eyed Mary's flowers are no more than an eighth of an inch across. They are common throughout the West, appearing as tiny blue dots underfoot soon after the snow melts. With magnification, you'll see that the flowers are attractive, with a lower lip made up of two turned-down, blue-tipped white petals. An unclassified form (above left), photographed on the Clearwater River, with flowers half an inch long, is sometimes seen in north-central Idaho. It seems to occupy a niche midway between the common blue-eyed Mary (above right) and an even larger species with flowers to three quarters of an inch long, *Collinsia grandiflora* Lindl., that grows in the Columbia Gorge, a plant that Lewis and Clark collected on April 17, 1806, early on their return journey.

Tweedy's snowlover (left, left center)
Chionophila tweedyi,
(Canby & Rose) L. F. Hend.

Tweedy's snowlover appears soon after the snowmelt. Half a dozen or so lavender-tinged flowers, their lips turned up at the end, bloom on one side of a stem that arises from a basal rosette of small oval leaves. The plant, related to the penstemons, was formerly classified as *Penstemon tweedyi*. It is found near treeline and lower in the mountains of Idaho and neighboring Montana. The name honors Frank Tweedy (1854–1937), a topographical engineer with the U.S. Geological Survey. *Chionophila* is from the Greek *xioni* meaning "snow" and *filos* for "friend." The inelegant name "toothbrush flower" has also been applied to the plant.

Taper-leaf penstemon
Penstemon attenuatus Douglas ex Lindl.

The taper-leaf penstemon (also known as sulphur penstemon for the color of one of its several varieties) is commonly seen in our mountains, sometimes in large numbers. It is characterized by its discrete, crowded flower clusters (verticillasters) and is sometimes confused with *Penstemon rydbergii* A. Nelson, a similar plant with well-defined flower clusters that grows at lower elevations. Four varieties of taper-leaf penstemon are recognized; all are found in Idaho (we believe this is var. *militaris* [Greene] Cronquist). The species is found mostly in the Northwest, although one variety also occurs in Wyoming.

Pale yellow penstemon
Penstemon confertus Douglas ex Lindl.

The pale yellow penstemon is found in the northern half of Idaho and in British Columbia, Alberta, Oregon and Montana. Its pale yellowish-white (ochroleucous) flowers distinguish the plant, as do its purple anther sacs and "bearded" stamen. The name *confertus* means "crowded," presumably for the flower clusters, although they are no more crowded than those of many other species. Penstemons may hybridize. This plant, for example, may cross with *Penstemon procerus* and produce a pink-flowered plant. The tendency for penstemons to hybridize suggests to botanists that the genus is actively evolving.

Dark blue penstemon
Penstemon cyaneus Pennell

The large-flowered, dark blue penstemon is a colorful plant that grows only in Idaho, Montana and Wyoming. A tendency for its flowers to be borne along one side of the stem distinguishes it from the similar Payette penstemon, *Penstemon payettensis* Nelson & J. F. MacBride, a slightly larger, showy species that grows with sagebrush in central Idaho and adjacent Oregon.

Fuzzy-tongue penstemon
Penstemon eriantherus
Pursh

The fuzzy-tongue penstemon bears a few large, wide, light purple flowers with prominent "guide-lines" within the throat. While *eriantherus* means "hairy anther," the markings and hairiness varies considerably among the five recognized varieties. The Idaho variety, var. *redactus* Pennell & D. D. Keck, is a low plant with a sparsely bearded yellow stamen.

Twin-leaved penstemon
Penstemon diphyllus Rydb.

The twin-leaved (*diphyllus*) penstemon is characterized by large pale-purple flowers whose upper lips are split for more than half their length, and by paired, toothed leaves ranged along the stem. A mountain plant, it grows in southern Idaho as well as in Washington and Montana.

Small-flowered penstemon
Penstemon procerus
Douglas ex Graham

This is the most common of Idaho's penstemons. It has small, narrow-tubed, densely arrayed flowers with lips that often close the mouth of the flower. The plants are common throughout the West, from the foothills to above treeline. Several varieties are recognized; ours is var. *procerus*.

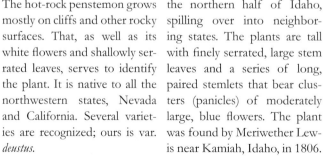

Hot-rock penstemon
Penstemon deustus
Douglas ex Lindl.

The hot-rock penstemon grows mostly on cliffs and other rocky surfaces. That, as well as its white flowers and shallowly serrated leaves, serves to identify the plant. It is native to all the northwestern states, Nevada and California. Several varieties are recognized; ours is var. *deustus*.

Wilcox's penstemon
Penstemon wilcoxii Rydb.

Wilcox's penstemon grows in the northern half of Idaho, spilling over into neighboring states. The plants are tall with finely serrated, large stem leaves and a series of long, paired stemlets that bear clusters (panicles) of moderately large, blue flowers. The plant was found by Meriwether Lewis near Kamiah, Idaho, in 1806.

Shrubby penstemon
Penstemon fruticosus
(Pursh) Greene

The shrubby penstemon is an attractive and common species found in the four northwestern states, British Columbia and Wyoming, growing from mid–elevations into the subalpine zone. Its stems are woody, and its leathery leaves are evergreen. When mature, the plants form shrubs that may grow to be quite large. Showy, light purple flowers are up to two inches long. Several varieties are recognized, identified mostly by the form of the leaves. Ours, illustrated here, with lanceolate, smooth-edged leaves, is var. *fruticosus,* the most common variety. Lewis and Clark found the shrubby penstemon on June 15, 1806, while on the Lolo Trail.

Mountain penstemon
Penstemon montanus Greene var.
montanus

This lovely penstemon is another woody species that grows in the mountains of Idaho, Montana, Wyoming and Utah. It is usually encountered by hikers, growing near treeline or higher. Like the shrubby penstemon, it forms well-demarcated shrubby clumps. Large violet flowers, woody stems, toothed leaves and the high elevation at which it grows, serve to identify the plant. A related variety, *Penstemon montanus* var. *idahoensis* (D. D. Keck) Cronquist, with mostly smooth-edged leaves, grows only in Idaho.

Scrophulariaceae (Plantaginaceae)

American brooklime
*Veronica americana**
Schwein. ex Benth.

Veronica flowers are irregular. The upper petal is large while the remaining three are distinctly smaller in size. While veronicas in general prefer moist places, some, like the American brooklime, are water plants that grow along the edges of small streams with their roots submerged. The term "brooklime" was derived from the Middle English "brok-lemok," for a closely related plant. (OED) Ours is easily identified by its growth preference, by its ovoid, finely serrated leaves and by its small, slightly irregular, blue flowers. The plant is found throughout the United States, except for several southern states. Other species are also common in Idaho—two more are shown here.

Cusick's speedwell
Veronica cusickii A. Gray

Cusick's speedwell is a subalpine plant that grows in the Pacific coastal states and in Idaho and Montana. Its small flowers are deep blue to violet, borne in a loose terminal cluster, or raceme. It may be identified by its long, yellow-tipped stamens, and even longer stigmas. William Conklin Cusick (1842–1922), for whom this plant was named, taught and ranched in Oregon. He found this namesake plant in the Blue Mountains of Oregon. It was described in 1878 by Harvard's professor of botany, Asa Gray, who classified many new plants sent to him from all over the world. The name "speedwell" has been in use for more than four centuries; it presumably is a testimonial to the plant's perceived medicinal value.

---

* The name "veronica" has been used for plants in this genus since at least the 1500s, although why they were so named is not known. The original veronica (*vera icon,* the "true icon"), preserved in St. Peter's in Rome, is the kerchief that a young woman (St. Veronica) used to wipe Christ's face on the road to Calvary. An impression of His face remained on the cloth.

American alpine speedwell
Veronica wormskjoldii
Roem. & J. A. Schultes

The American alpine speedwell, *Veronica wormskjoldii*, is an attractive, small, sub-alpine flower found in the mountains of the West and across the northern part of the continent to Greenland. Its stem is hairy and its flowers lack the long anthers and stigmas of Cusick's speedwell, shown on the preceding page. Morten Wormskjold (1783–1845) was a Danish botanist who, in 1813, led an expedition to Greenland where he collected many plants, including this one. Later, he accompanied the Kotzebue expedition to the North Pacific in the company of botanists Adelbert von Chamisso and Johann Friedrick Gustav von Eschscholtz; both are mentioned elsewhere in this book.

Mountain kittentoes
Synthyris missurica (Raf.) Pennel

Mountain kittentoes favor shaded, moist woods. Their tiny, four-petaled flowers are similar to those of the veronicas, and for a time the plant was classified as *Veronica missurica* (with ongoing taxonomic study, it may become so again). The small leafy bracts below the flower cluster help to identify this and several related species, as do its broad, scalloped basal leaves. This species is found only in the Northwest and in California. The plant shown here was photographed in deep woods along US Highway 12 adjacent to the Clearwater River, close to where Meriwether Lewis collected the plant on June 26, 1806, on the expedition's return journey.

Butter-and-egg plant, *Linaria vulgaris* Hill (top)
Dalmatian toadflax, *Linaria dalmatica* (L.) P. Mill. (center)

When one first sees the common butter-and-egg plant, its clustered butter-yellow and egg-yolk-orange flowers immediately identify the plant. The butter-and-egg plant and the related Dalmatian toadflax, *Linaria dalmatica* (center) are similiar both in appearance and behavior. The plants were introduced from Europe as ornamentals during the 1800s, with no concern about the effect that they might have on our native flora. Both now grow in Idaho as high as the subalpine zone, sometimes forming dense spreading yellow patches of plants that crowd out other species. Both are ranked as serious weeds throughout the United States.

The two species do differ a bit. The Dalmatian toadflax has a looser flower cluster, the flowers' basal spurs are sharper, the leaves are wider and they clasp the stem—the latter is the most consistent identifying feature. Toadflaxes in general have three-petaled upper and two-petaled lower "lips" that meet to resemble a toad's mouth. This and flax-like leaves explain the name "toadflax." *Linaria* was derived from *Linum,* the generic name for flax.

Common mullein, *Verbascum thapsus* L. (bottom)

Because of the taxonomic changes noted in the introduction to this chapter, the common mullein, *Verbascum thapsus,* and *Limosella acaulis* (page 200) are the only true Scrophulariaceae shown on these pages. The mullein—a serious weed—has spread far and wide since it was introduced from Eurasia and is at home in every state and most Canadian provinces. It grows on dry disturbed ground, alongside roads, in pasture land, near construction sites and even away from settled places in our mountains. Large dusty-green leaves, long stems and dense clusters of irregular, yellow, five-petaled flowers make the common mullein recognizable immediately. The plants are biennial, forming small furry-leaved rosettes the first year and the familiar long-stalked plants the next. They produce large numbers of hardy seeds—seeds a century old are said to have germinated. Formerly, the plant's leaves were used medicinally for tea and poultices, and smoked to treat asthma. Dried mullein stalks, saturated with grease or wax, were used as torches. *Verbascum* is an ancient name for common mulleins; the word "mullein" is from the Latin *mollis* meaning "soft" for the velvety leaves. *Thapsus* is the name of a city in Sicily.

Nightshade Family (Solanaceae)

The Solanaceae, or Nightshade, family is made up of ninety-eight genera and 2,715 species. It includes some of our most important food plants: potatoes, peppers, tomatoes, eggplant, etc. Tobacco is also a Solanaceae. Many members of this family produce poisonous alkaloids including nicotine and atropine-related substances. Some of the latter are therapeutically important: atropine (belladonna), scopolamine, hyosciamine and other congeners.

Various other Solanaceae are favorite garden ornamentals, including petunias, Japanese lanterns and others. While three dozen or so genera are native to North America—mostly as semi-tropical plants—only a few are found in Idaho. These include species of *Datura* (Jimson-weeds), *Physalis* (ground-cherries), *Solanum* (nightshades and related plants) and *Nicotiana* (tobacco); members of the last two genera may be encountered in Idaho growing as high as the montane zone.

Two nonnative Solanaceae are also included here. Both are Eurasian imports now at home throughout North America. They too may be seen growing at fairly high elevations.

The family takes its name from *solanum*, an old Latin name that means "comforter," probably from the same root as "solace," and reflects the sedative effects of various alkaloids found in plants in this family.

Coyote tobacco
Nicotiana attenuata Torr. ex S. Wats.

Coyote tobacco is a tall, somewhat sticky and rather wispy (*attenuata*) annual plant that occasionally appears in dry, open places along our trails in midsummer. Its flowers, similar to those of garden varieties of *Nicotiana,* make it easy to identify as a tobacco plant at first sight. The plant's leaves, especially the large basal leaves, smell strongly of tobacco. Coyote tobacco ranges throughout the West, from British Columbia to Texas. The plant was important ceremonially to Native Americans, who used the leaves as others use those of cultivated tobacco (chiefly *Nicotiana tabacum* L.) today.

Cutleaf nightshade
Solanum triflorum Nutt.

The cutleaf nightshade bears its flowers and fruit in clusters of three, hence the species name *triflorum*. It is also known as wild tomato; its rather furry leaves and orange berries do, in fact, suggest a tomato plant (also in the Solanaceae family). The fruit is said to be edible, although we would be reluctant to try any wild fruit in this family (the plant is, in fact, included in several lists of poisonous plants). It is found in most of the United States and Canadian provinces, and on other continents as well. Cutleaf nightshade is classified as a native plant, but because it occurs so widely some believe it may be an import. It occurs but is uncommon in montane Idaho.

Climbing nightshade
Solanum dulcamara L.

The climbing nightshade usually grows on disturbed ground. While considered a weed, it is not a particularly agressive one. The flowers' reflexed, deep purple petals and joined, beak-like yellow anthers serve to identify the plant. Although the leaves and unripe fruit are poisonous, the ripe red berries are toxic only if eaten in large quantities. The berries are bitter at first, but then leave a sweet aftertaste, explaining the species name, *dulcamara*, Latin for "bittersweet." The climbing nightshade, like the henbane below, is an imported Eurasian plant.

Black henbane
Hyoscyamus niger L.

Black henbane, given its striking appearance, is recognizable at first glance. An Old World plant, it is now widespread throughout North America. The plant is considered a noxious weed in Idaho and in several other states. Henbanes contain generous amount of the alkaloid hyoscine (scopolamine), used for millennia as a sedative, and for its atropine-like properties. Scopolamine has been used in obstetrics to induce "twilight sleep," as an adjunct to anesthesia and—reputedly—for extracting confessions.

Valerian Family (Valerianaceae)

The word "valerian" possibly was derived from the Latin word *valere* meaning "to be strong," or "healthy"—a tribute to the plants' supposed medicinal worth. The word has been in use for a long time; the OED's first citation is to Chaucer, who mentioned the plant ca. 1386, and presumably it was named even earlier. The family is not a large one, for it includes only about nine genera and 400 species. While most grow in the north temperate zone, a few are found in the South American Andes. The common European garden heliotrope, *Valeriana officinalis* L. (its species name implies "of the apothecary shop"), resembles the Sitka valerian, a plant common in Idaho's mountains. The roots of the European plant contain a pharmacologically active principle, valerianic acid, said to act as a sedative and sleep aid; presumably the same substance occurs in our plants. Valerian flowers are small, mostly white, and clustered, with five petals and three stamens. There are usually paired opposing leaves below the inflorescence. Toward the base, the leaves are whorled and often compound. Plants belonging to three different genera of Valerianaceae grow in the Northwest, but the four species of genus *Valeriana* shown here are the only members of the family that we have seen in our mountains.

Edible valerian
Valeriana edulis Nutt. ex Torr. & A. Gray var. *edulis*

The edible valerian, or tobacco root, is a conspicuously leafy perennial plant growing from a substantial taproot, found most commonly in recently moist places. It bears several crowded, creamy-white flower clusters on each of several thick stems. The species name, *edulis,* means "edible," referring here, apparently, to the plant's bulky roots. Although these were gathered, eaten and apparently enjoyed by Native Americans, they are, by report, "the most horrid food ever ingested." (Harrington, HD, *Edible Native Plants of the Rocky Mountains*). Valerian roots in general have a strong and rather unpleasant odor that some have compared to smelly socks. Cats are attracted to the dried roots; they are said to have the same effect as catnip. Another variety, *Valeriana edulis* var. *ciliata* (Torr. & Gray) Cronquist, grows in several midwestern states north to Canada, but ours (var. *edulis)* is found only in the western mountain states and as far south as northern Mexico. This plant was first collected as a scientific specimen by Thomas Nuttall near today's Walla Walla, Washington, and described by Torrey and Gray in 1858.

Unlike many plants with reputed medicinal properties, valerianic acid (extracted from the roots of the Eurasian *Valeriana officinalis*, but probably present in other species as well) has been shown to function, for some people at least, as an effective sedative.

Sharp-leaf valerian
Valeriana acutiloba Rydb.
var. *pubicarpa* (Rydb.) Cronquist

The sharp-leaf valerian has softly bristled seeds, explaining another common name, downy-fruited valerian. The name *acutiloba* refers to the plant's pointed leaves made up of three or more leaflets. A high-altitude plant, it grows to treeline in the mountains of all the western states, blooming early in the summer while north-facing slopes are still snow covered. The small, sometimes pink-tinged flowers bloom centripetally, as do those in the Vervain family (Verbenaceae), explaining another common name, mountain verbena. The name "cordilleran valerian" has also been suggested as a common name for this plant.

Northern valerian
Valeriana dioica L. var. *sylvatica* (Richardson) S. Watson

The northern, or marsh, valerian, is—like most members of the genus—stout-stemmed. It is a low plant, usually found growing in moist mountain meadows or on ground wet from the recent snowmelt, at mid- to subalpine elevations. The species name, *dioica,* from the Greek, means "two houses," signifying that there are both male and female plants. While not showy, these are not unattractive plants and are said to do well in wildflower gardens. The northern valerian is at home in the mountains of most of our western states and Canadian provinces. Because it is a circumboreal species, our plant is classified as var. *sylvatica*, to distinguish it from the slightly different Eurasian variety, var. *dioica*.

Sitka valerian, *Valeriana sitchensis* Bong.

The Sitka valerian, or mountain heliotrope, common in Idaho, prefers the open shade of evergreen forests. The plants bloom in late spring at montane elevations, and later in the subalpine zone. Usually one tight cluster of small white flowers is borne on each stem. Groups of opposing lanceolate leaves appear at intervals along the stem, growing so closely together as to appear whorled. They do not show well in this crowded image, but the plant's compound leaves typically are made up of pointed leaflets—a large one and two or more smaller paired lobes. The plants spread by their roots, so commonly several to many are found growing together. Three stamens and a pistil extend well beyond the flowers' five petals, giving the flowerhead a feathery look. The species name, *sitchensis*, refers to the Russian settlement of Sitka, on Baranof Island, where the plant was first collected by Karl Heinrich Mertens (1796–1830).

Violet Family (Violaceae)

Violets, wild and cultivated, have been treasured for millennia. (African violets, *Saintpaulia* spp., are not in the Violet family at all, but are members of another family, Gesneriaceae.) The family Violaceae is made up of twenty-three genera and 900 species. It is widely distributed; most are found in the north temperate zone. About sixty species are native to the United States, and about half a dozen species grow in our mountains. The name "violet" was derived from the Latin word *viola,* used for a sweet-smelling flower, possibly the European violet, *Viola odorata* L., a small pansy-like flower known to Europeans as "hearts-ease," and to us as "Johnny-jump-up"; the plant is believed to be an ancestor of today's cultivated pansies.

Violets typically have five sepals and five separated petals. The flowers are irregular but bilaterally symmetrical, with a large, sometimes bearded lower petal. A nectar-containing sac or spur attached to the lower petal may extend backward behind the flower. Most violets prefer shade and moisture, but some, like our goosefoot yellow violet, *Viola purpurea,* flourish on dry ground. Violets tend to be spring bloomers, recognizing that "spring" for flowers represents a condition rather than a date, and varies with altitude. Some species may flower again in the fall if the weather is mild. The family has little economic importance beyond its many popular pansy cultivars.

Hooked violet, *Viola adunca* Sm.

If, in hiking along our muddy trails and through spring-moist meadows anywhere in Idaho, you see blue violets, they are most likely *Viola adunca,* the hooked violet ("early blue violet" is another common name). Although it doesn't show well in our image, if you look closely at a flower in the field, you will see that it has a prominent upward spur (derived from the lower central petal) extending backward behind the flower. The hooked violet grows throughout the West and across northern North America. Our plant is var. *adunca;* two other varieties grow only in California. The species name, *adunca,* means "hook" in Latin.

Goosefoot yellow violet, *Viola purpurea* Kellogg
var. *venosa* (S. Watson) M. S. Baker & J. C. Clausen

The plant shown here grows in the coastal states, east to Wyoming, Utah and Colorado. It is also found, rarely, in British Columbia and Alberta. A variable plant, it has had at least ten names over the years, as well as forms now recognized as varieties. We believe our Idaho plant is best classified as above. The goosefoot violet (so named for the pattern of veins in the leaves) grows on gravelly slopes, often in the company of sagebrush. It is one of the earliest spring flowers, appearing immediately after snowmelt at ever higher elevations, as high as the subalpine zone.

Pioneer violet, *Viola glabella* Nutt.

The pioneer, or stream, violet grows in northern California, throughout the Northwest, north to Alaska and in eastern Asia. It appears in great numbers in northern Idaho, growing on the moist ground of deep woods and along streambanks. It may be identified by its bright yellow flowers marked with black pencilling and by its heart- or kidney-shaped leaves. The plant was collected first by Thomas Nuttall in the coastal mountains of Oregon during the spring of 1835.

Valley violet, *Viola vallicola* A. Nelson

Previously known as Nuttall's violet (a plant confined to the Great Plains), the valley violet is distinguished by its large lanceolate leaves. These may be smooth-edged, or sometimes have shallow teeth as in the illustration. The plant grows throughout the West in both the United States and Canada, ranging as far east as Kansas and the Dakotas. It grows from mid- to high elevations along streambanks and on moist ground.

Small white violet, *Viola macloskeyi* F. E. Lloyd

The common name "small white violet" suits this plant, for its purple-marked flowers are the smallest of any of our violets' flowers. The plant grows from mid-elevations to high in our mountains. Macloskey's violet is widely distributed, occurring along both coasts, in our northern states and in the Canadian provinces (although rare in Alberta and Saskatchewan). Two varieties are recognized: one with scalloped leaves, var. *pallens* (Hanks ex Ging.) C. L. Hitch., and the other with smooth-edged leaves, var. *macloskeyi*. The latter is said not to occur in Idaho, although those shown here (photographed near Galena Summit) certainly fit the description of that variety. George Macloskey (1834–1920), for whom the species was named, was born in Ireland. He held degrees in both theology and natural sciences, becoming professor of biology at Princeton University in 1874, a chair he held until 1906.

Iris Family (Iridaceae)

The Iridaceae, a family of colorful flowers, was named for Iris, the Greek goddess of the rainbow. It is made up of sixty genera and 1,845 species. Many Iridaceae—*Gladiolus*, *Crocus*, *Freesia*, *Iris* and others—are found wherever flowers are cultivated; ornamental plants are the family's main economic importance. Most are perennials, spreading by bulbs and root-like underground stems (rhizomes) and by seeds. A non-botanist looking at an iris's showy flower may have difficulty identifying the parts, but usually there are three bent-back petal-like sepals ("falls") and three erect petals ("standards") joined at their bases to form a swollen floral tube—a family characteristic. There are three hidden stamens and three stigmas that—confusingly—in some species take the shape and color of a petal. Leaves are equitant, meaning that the sides of the longitudinally folded leaves ride on either side of adjacent stems. The parallel-veined leaves are otherwise typical of monocotyledonous plants in general. Only three wild irises are found in Idaho. One, not shown here, is *Iris versicolor*, the harlequin iris, common in the northeastern states and provinces. A disjunct population—it is a rare plant in Idaho—grows at the head of Priest Lake in Idaho's panhandle. The other two, the western blue flag (*Iris missouriensis*) and the yellow flag (*Iris pseudacorus*—an import), are shown below and on the following page in company with another member of the family, our native blue-eyed grass, *Sisyrinchium idahoense*.

Western blue flag
Iris missouriensis Nutt.

The western blue flag grows in most western states and as far east as Minnesota. It is easily identified, for it looks like a garden iris on a diet. You will find it blooming, usually for only a short time, in late spring. It grows as high as the subalpine zone, always in moist places. Its color varies from light to dark blue according to location. Meriwether Lewis collected the plant on July 5, 1806, while ascending the Blackfoot River in Montana on his return trip. Only fragments of his specimen survived, so botanist Frederick Pursh could not publish a description. It was again collected by Nathaniel Wyeth in 1833 and described by his friend, Thomas Nuttall. Not everyone is enchanted by this attractive plant, and it is classified as a weed in California and Nevada.

Yellow flag, *Iris pseudacorus* L.

The yellow flag was imported from the Old World as an ornamental water plant. It has made itself at home in the New World, forming spreading clumps in the shallow water of ponds and streams (the plant shown was photographed in Blaine County's Silver Creek, a prime trout water). It is immediately identifiable by its pale to deep yellow flowers and its growth habit. Like the western blue flag, a native plant, the exotic yellow flag is also classified as a noxious weed in Nevada and California. The *-acora* in the species name was one bestowed by the Greek nauralist Theophrastus (372–287 BC) on an iris.

Idaho blue-eyed grass
***Sisyrinchium idahoense* E. P. Bricknell**
var. *occidentale* (E. P. Bricknell) D. M. Hend.

The blue-eyed grasses are Iridaceae, with narrow, grass-like basal leaves. A single flower on the end of a naked stem bears three matching petals distinguished by their sharply pointed tips and three blue petaloid sepals with rounded ends and no pointed tips. Four varieties (formerly species) of *Sisyrinchium idahoense* are recognized (two grow in Idaho). In the past several sisyrinchiums were lumped into one species, *Sisyrinchium occidentale,* retaining their former names as varieties. Blue-eyed grasses favor moist meadows, often growing in the company of western blue flags. They are said to do well in ornamental gardens. The species name was derived from the name *Sisyrinchus,* also used by Theophrastus for an iris-like plant.

Lily Family (Liliaceae)

The Lily family (including many newly recognized and related families) consists of almost 300 genera and 5,000 species. It has always been considered a taxonomic catch-all family, one made up of many diverse genera with a few common characteristics. While most have bulbs or corms (thick bulb-like stems) that divide belowground, some spread by rhizomes (creeping underground stems). All have the parallel-veined leaves characteristic of monocotyledonous plants in general. Most also are perennial, herbaceous (non-woody) plants whose flowers have three sepals and three petals (often modified into six similar tepals). Despite these common features, there is good reason—based on morphologic and molecular differences—to divide the Liliaceae into a number of smaller families. There is less agreement among taxonomists, however, as to how this should be done. It seems reasonable, at least for our purposes, to retain the older classification while indicating those new families that are now generally accepted. These include:

Agavaceae (Agave family) *Camassia*

Alliaceae (Onion family) *Allium*

Calochortaceae (Mariposa lily family) *Calochortus*

Liliaceae (Lily family) *Clintonia, Prosartes* (formerly *Disporum*), *Erythronium, Fritillaria, Lloydia* and *Streptopus*

Melanthaceae (Bunchflower family) *Trillium, Stenanthium, Veratrum, Xerophyllum, Toxicoscordion* and *Anticlea* (The plants in the last two genera were, until recently, assigned to *Zigadenus*, a broadly defined genus.)

Ruscaceae (Butcher's broom family) *Maianthemum*

Themidaceae (Cluster lily family) *Triteleia*

Aase's onion, *Allium aaseae* Ownbey

Until recently Aase's onion was thought to be a rare plant, growing only in the Boise foothills. More recently, however, specimens have been found farther west, close to the Idaho-Oregon border. These lovely little plants are one of the most atractive of all our wild onions. They bloom early in the spring on south facing sandy slopes, sometimes in fairly large numbers. Aase's onion is still considered a rare species, and steps have been taken to protect the plants wherever they grow. The species name honors botanist Hannah Aase (1883–1980), an academic botanist on the faculty of the University of Washington from 1914 to 1949. Dr. Aase's interests included plant genetics of cereal grasses as well as the family Alliaceae (the Onion family). She wrote (with co-worker Marion Ownbey, also of the University of Washington) a survey on the genus *Allium,* leading to the use of her name for the specific epithet of this wild onion.

219

Hooker's onion
Allium acuminatum Hook.

Hooker's onion (also known as taper-tip onion) was collected by Archibald Menzies, ship's surgeon and botanist of the Vancouver Expedition, on to-day's Vancouver Island in the 1790s. It was later classified by William Jackson Hooker, then professor of botany in Glasgow and later director of England's Kew Gardens. His name is still associated with the plant. The word "acuminate" means "to taper to a sharp point," from the shape of the tepals. Each of our wild onions has its own ecological niche. This one prefers dry ground, growing to moderately high elevations and blooming in summer after other flowers have gone by. Said to be the commonest wild onion in the Northwest, it is found throughout the mountain West. The pink to purple flowers are borne in a loose umbel atop a long, thin, leafless stem. The leaves are rather wispy and inconspicuous, and have usually dried up by the time the flowers appear.

Short-styled onion
Allium brevistylum S. Watson

The short-styled onion (the common name is a translation of the scientific species name) is a tall plant whose flowers are borne in a loose umbel. The flowers do not open widely, and the short style, from which the plant takes its names, can't be seen easily, as it can with many other onions. The plant's stems are much longer than its leaves. It prefers streambanks and other moist locations, growing at least as high as treeline. At lower altitudes the leaves are bright green as shown here; at treeline they take on a reddish hue. The plant grows only in Idaho, Wyoming, Colorado and Utah.

Brandegee's onion
Allium brandegeei S. Watson

Brandegee's onion grows on gravelly hillsides at all elevations, appearing soon after the snowmelt. Often there are so many as to almost turn the ground white with their flowers. It is a species often seen in our mountains, growing from northeastern Oregon, across southern and central Idaho to Utah. Disjunct populations also occur in neighboring states. Each flower has six tepals and each tepal has a dark rib, often more obvious on the outer side. The flowers, usually white, are occasionally pink. Its leaves and the bulbs taste much like garden onions (*Allium cepa* L.) and were used in the same way by early settlers. Animals also appreciate the plants; the leaves are often cropped—as in the illustration—by grazing deer. The species name refers to Townshend Stith Brandegee (1843–1925), an American civil engineer, botanist and plant collector, who collected the plant in the Elk Mountains of Colorado.

Geyer's onion
Allium geyeri S. Watson var. *tenerum* M. E. Jones

Geyer's onion is another common Rocky Mountain onion that grows, often in large numbers, along mountain streams and in moist meadows, sometimes in company with the short-styled onion shown on the previous page. Two varieties occur in Idaho, var. *geyeri* in which the heads lack small bulbs (bulbils), and var. *tenerum* (*tenerum* means "tender" or "soft") in which many, or all, of the flowers are replaced by bulbils. A tall white-flowered onion, the bulbils, when present, make identification easy. The species name honors Charles A. Geyer (1809–1853), a German botanist and plant-hunter who was hired to collect plants in the western United States. In 1843, he traveled across "upper Oregon" (today's western Montana, northern Idaho and eastern Washington) gathering thousands of plant specimens; a dozen or so now bear his name.

Chive
Allium schoenoprasum L.

Wild chives have a rather spotty distribution across the northern states and Canadian provinces, west into Asia and Europe. Our wild plant appears identical to the garden chive (as it should; it is the same species). The plant is easily identified—most everyone knows what chives looks like. Even lacking its light purple flowerhead, the typical appearance of its hollow leaves and its taste should clinch its identification. The species name, *schoenoprasum*, was derived from two Greek words meaning "rush" and "leek." Strictly, "chive" (an Old English word) should refer to the plant, and "chives" to its culinary use.

Allium simillimum L. F. Hend.

The pink onion shown on the left was photographed on a mountain trail immediately north of the Sun Valley resort. We included this illustration in the first edition of this book as *Allium aaseae,* a similar, but uncommon, pink-flowered species (described on page 219). Several botanists acquainted with the latter plant disputed the classification. Since then, a specimen has been examined by three knowledgeable botanists who believe it to be a form of *Allium simillimum,* a (usually) white-tepaled plant found in Idaho and (rarely) in Montana. The name *simillimum,* from the Latin, means "similar to," possibly to another species of wild onion. There seems to be no established common name for the plant.

Tolmie's onion
Allium tolmiei Baker

Tolmie's onion is an attractive plant, although not a particularly common one. It is found in the western part of north-central Idaho and in the coastal states from northeastern California to northeastern Washington. The most common variety, shown here, grows throughout that range. Two other localized varieties that differ slightly in morphology are also recognized. Tolmie's onion grows on dry, gravelly ground, rather than in the moist areas preferred by many other members of the Onion family. It may be identified by its pink flowerheads and especially by its leaves. These are notably wide, flat and sickle-shaped and are quite long compared to the short-stemmed flowerheads. The anthers are longer (up to twice as long) than the tepals in this variety.

William Frazer Tolmie (1812–1886) was a surgeon with the Hudson's Bay Company. He was given a plant collection, including this onion, by a friend. Tolmie sent the collection to William Jackson Hooker in England. Subsequently Tolmie's name was attached to the plant. Tolmie explored the country around Mount Rainier in 1833 and apparently hoped—or possibly even attempted—to climb the mountain. (Later, others attempted the climb, but the first successful ascent was not until 1870.)

Common camas
Camassia quamash (Pursh) Greene

The Indian word *quamash* is the source of both generic and species names for this plant; it also supplied the common name, anglicized as "camas." Explorers were amazed when they saw fields of camas in bloom. Meriwether Lewis wrote in June 1806 that the plants stretched like "lakes of fine clear water."

In Idaho and adjacent states, the flowers still form spectacular "lakes" each spring on water-soaked fields. Their bulbs are the size of a small onion and are palatable both raw and cooked. They store well across the seasons, which made them a prime food source for Native Americans. The members of the Lewis and Clark Expedition gratefully ate camas roots in the camp of Nez Perce Indians following the perilous crossing of the Bitterroot Mountains in the autumn of 1805. Many of the explorers became sick. Meriwether Lewis almost died, and blamed the roots for his illness (more likely their gastroenteritis was derived from tainted salmon that the Indians also gave them). Nevertheless, he gathered a specimen of camas on the Weippe Prairie in June 1806, while waiting to cross the Lolo Trail on the return journey. The distinctive flowers are large with six similar tepals and bright yellow, parenthesis-shaped anthers. Flower color ranges considerably from one location to another, from almost white, to shades of blue, to the blue-gray color shown here, to purple. The illustration below shows a "lake" on Camas Prairie near Fairfield, one of several "Camas Prairies" in Idaho.

Feathery false Solomon's seal
Maianthemum racemosum (L.) Link
var. *amplexicaule* (Nutt.) Dorn

The feathery false Solomon's seal (also known as western false Solomon's plume and wild lily of the valley) is a common plant found in much of North America. Until recently the two plants shown on this page were in the genus *Smilacina* and are still so identified in many guidebooks. On the basis of recent studies, however, they have been placed in the Butcher's broom family (Ruscaceae) and in the genus *Maianthemum*. (The latter name was derived from two Greek words for "May" and "flower.") The maianthemums, and many similar lily-like plants, commonly grow in forest shade and other moist situations. Deep green, parallel-veined leaves and a spray of small-petaled, yellow-anthered, white flowers make identification easy. Green berries turn red as they ripen. Botanists recognize two varieties; ours, var. *amplexicaule* (the word means "stem-clasping" for the leaves) is a western plant. The other, var. *racemosum,* grows in the east; the two varieties are separated by the Great Plains.

Starry false Solomon's seal
Maianthemum stellatum (L.) Link

As is common with widely distributed plants, the starry false Solomon's seal has many common names. These include star-flowered (or western) Solomon's plume, wild (or false) lily of the valley and others. The plant is quite similar in appearance to the feathery false Solomon's seal shown above, differing in that its leaves are narrower, its tepals are larger, and the flowers are star-shaped (the species name, *stellatum,* was derived from the Latin word *stella* for "star"). The starry false Solomon's seal grows in all the states and provinces of North America, except for Texas and the southern seaboard states.

Large-flowered triplet-lily
Triteleia grandiflora Lindl.

The large-flowered triplet-lily is an attractive flower, now classified in the Cluster lily family (Themidaceae). It blooms early in dry meadows to mid-elevations. Its flowers have three outer and three inner tepals; the latter have ruffled edges. Flower color ranges from white with blue markings (var. *grandiflora*) to blue with darker lines (var. *howellii* [S. Watson] Hoover). The species grows in the northwestern states, British Columbia, Utah and—uncommonly—in neighboring states.

Originally collected by David Douglas, the plant was until recently classified as a *Brodiaea,* a genus with three-anthered flowers. The British botanist John Lindley (1799–1865) published a description of our plant in 1830, noting that its flowers have six anthers (as Douglas had noted), and placed it in a new genus, *Triteleia*. Recent studies have shown that his classification was correct—although it is still often listed as "Douglas's brodiaea." Meriwether Lewis collected the plant while ascending the Columbia River on April 17, 1806. Frederick Pursh noted that Lewis's flower had six anthers. Unfortunately, he classified it as a *Brodiaea*, thus missing a chance to describe and name a new genus.

Glacier lily
Erythronium grandiflorum Pursh var. *grandiflorum*

Glacier lilies are found throughout the northwestern states and Canadian provinces and east into the Rocky Mountain states. As with other well-loved and widely distributed plants, this one has many common names: glacier lily, avalanche lily, trout lily, dog-tooth violet, fawn lily, adder's tongue and others. It is a common plant in parts of Idaho. Although we have not seen it growing south of Lemhi Pass (thirty miles southeast of Salmon), it does grow farther south in the mountains of the southeastern part of the state. Lewis and Clark collected the glacier lily near their Kooskooskee (Clearwater) River campsite on May 8, 1806, and later on the Lolo Trail. A related white variety, var. *candidum* Piper, grows only in northern Idaho, Washington and Montana. (This plant is shown out of sequence. It is a Liliaceae, which starts on page 227.)

Sharon Huff

Sego lily, *Calochortus nuttallii* Torr. & Gray

At first glance there is little difference in appearance between the sego lily shown on the left and the white mariposa lily shown below. Look more closely and you will see that in the sego lily the purple spots on the petals are more linear and lower, bordering a yellow gland. The sepals between the petals are thicker, shorter and different lengths, and the petals are a different shape. Finally, the sego lily has a tangle of hairs, the "beard," at the base of the petals. *Calochortus nuttallii* is the Utah state flower, the plant that kept Mormon settlers alive during difficult times, or so the story goes. In earlier times the plant was said to occur in Idaho only in the southeastern corner, adjacent to Utah. We now know that it occurs in a more or less patchy fashion throughout Idaho and the interior states of the West, east to the Dakotas and south to New Mexico, blooming from late spring on.

Three spot mariposa lily
***Calochortus apiculatus* Baker**

Calochortus apiculatus prefers moist woods. Its distribution makes it look as if the plant originated along the Idaho-Canadian border and then spread centrifugally within Idaho's panhandle, to the nearby corners of Alberta and British Columbia and the adjacent corners of Washington and Montana. Its sepals are a bit shorter than the petals. There are small purple spots adjacent to circular yellow glands at the base of the petals, giving off a sparse beard and sharply pointed anthers. Each plant is usually single-flowered, but may bear more. This mariposa lily is not hard to identify; it grows in Idaho only in the panhandle, where it blooms early in spring.

White mariposa lily
***Calochortus eurycarpus* S. Watson**

The white mariposa lily (also known as the star tulip) is identified by the deep purple spot at the base of each petal. Farther in, there is a bristle-lined yellow to green gland. The petals are delta-shaped; the sepals are only a little longer than the petals. The species grows in Idaho and the six surrounding states. Great numbers appear in early to midsummer in our mountains, growing almost to treeline. The plants are known regionally as "sego lilies," although the true sego lily is *Calochortus nuttallii*, the Utah state flower, occurring farther south. This plant's species name means "broad seed," from the Greek.

Elegant mariposa lily
Calochortus elegans

The elegant mariposa lily's small flower only measures about an inch across. The plant was first collected by Meriwether Lewis on May 17, 1806, near today's Kamiah, Idaho. He brought the dried specimen back to the United States where botanist Frederick Pursh recognized that the plant was new to science. Pursh coined the name *Calochortus* (from two Greek words meaning "beautiful plant") for a new genus, giving it the scientific name that it has today. There are several varieties of *Calochortus elegans* including var. *nanus* Ownbey (southeastern Oregon and northern California) and var. *selwayensis* Ownbey (Idaho and Montana). Ours is var. *elegans*, the variety that Lewis collected, and this one was photographed close to where the explorer found his.

Sagebrush mariposa lily
Calochortus macrocarpus Douglas

The sagebrush mariposa lily (also known as green-banded mariposa lily) is a foothills plant that blooms in midsummer on dry, sagebrush-covered slopes south of the Clearwater River drainage. The species name means "long fruit" for the shape of its seed. The three petals and three sepals are both the same color, ranging from pale pink to lavender. Its narrow sepals are noticeably longer than the pointed ovoid petals. A green band (top image) sometimes runs down the back of each petal. These lovely plants do not do well in cultivation; enjoy them where they grow!

Columbia lily
Lilium columbianum Leichtin

The Columbia lily is a tall, multi-flowered plant whose flowers look as if they should be in an ornamental garden. It is found in Idaho only in Boundary and Bonner Counties at the top of the state's panhandle. Its range includes the coastal states and British Columbia, as well as Montana. The plant grows in gardens under a variety of names.

Fairy bells
Prosartes trachycarpa S. Watson

Fairy bells grow in the shade of evergreen forests at higher altitudes, often in the company of species of *Maianthemum* (false Solomon seals) and *Berberis repens* (creeping Oregon grape). The plants are easily identified by their wide, parallel-veined leaves and long-stalked, lily-like, paired flowers (above). The ovaries mature into bright red, irregularly shaped berries (below). The berries are said to be edible, but tasteless. Until recently this plant was classified as *Disporum trachycarpum*. Now it has a restored generic name, *Prosartes,* apparently derived from a Greek word meaning "attached," referring to the ovules in the fruit. Our species grows in many western states and Canadian provinces, with disjunct populations in the East.

Clintonia
Clintonia uniflora
(Menzies ex Schultes) Kunth

Clintonias go by many common names including beadlily, blue-bead lily, bride's bonnet, queen's cup and simply clintonia. They are shade-loving, montane to subalpine plants with a six-tepaled, single white flower (*uniflora*). Their stems arise from a cluster of two or more wide, rather leathery leaves. Each plant bears one blue berry (below). They grow in coastal mountain ranges from California to Alaska, and inland to Idaho and Montana. The genus name, *Clintonia*, honors amateur naturalist DeWitt Clinton (1769–1828), also known for his political career as a senator, presidential candidate, promoter of steam navigation and governor of New York.

Yellow bells, *Fritillaria pudica* (Pursh) Spreng. (left and right)

Yellow bells (also known as yellow fritillaries and yellow mission-bells) bloom in April at lower elevations and well into June higher up, as single plants or in small groups. Narrow, twisted leaves appear first, followed by the flowers. These have six tepals arranged in two rows. Its tepals are yellow-orange with a deep red basal band. The plant is native to most of the far western states. Its species name, *pudica*, means "modest" for the shy, downward- facing flowers. *Fritillaria*, in turn, was derived from a Latin word for "dice-box," presumably for the seed-containing capsule (right). Lewis and Clark collected the plant on May 8, 1806, on the Clearwater River.

Checker lily (left)
***Fritillaria affinis* (Schultes) Sealy**
Chocolate lily (right)
***Fritillaria atropurpurea* Nutt.**

These two plants are clearly related, although their ranges differ. Both bloom about a month later than the yellow bells shown above. The chocolate lily grows in the southern half of Idaho and in states to the south. The checker lily grows farther north in Idaho and in the northwestern states to British Columbia. Neither plant is particularly common and finding one is a cause for comment. Fritillaries are sometimes called "rice-roots" for a myriad of tiny bulblets attached to a main bulb. Lewis and Clark collected *Fritillaria affinis* on April 10, 1806, while ascending the Columbia River.

Common alp-lily
Lloydia serotina (L.) Reichenb. var. *serotina*

The alp-lily (also commonly known as lloydia) is found in the mountains of many of the western states (although rare in Oregon), north to Alaska and in the mountains of Eurasia. It is an unprepossessing little plant, a true alpine species. A few narrow, inconspicuous leaves and yellow-anthered, white, six-tepaled flowers less than a quarter of an inch across identify the plant as a lloydia. The plant was named for Edward Lloyd (1660–1709), prominent Welsh author, naturalist and curator of the Ashmolean Library of Oxford University, who discovered this plant growing in Wales and recognized that true alpine flora grew above treeline in the Welsh mountains. The scientific species name, *serotina,* means "late" or "autumnal," a strange choice, for the flowers bloom while there is still snow on surrounding slopes. A rare yellow variety, var. *flava* (Calder & Roy L. Taylor) B. Boivin, occurs in British Columbia.

Clasping-leaf twisted-stalk
Streptopus amplexifolius (L.) DC.

The twisted-stalk's stem zigzags from one leaf node to the next as reflected in its generic name, *Streptopus* (from the Greek meaning "twisted foot"), and its stemless leaves clasp the main stem. Dependent white flowers have reflexed (turned back) tepals. Its oval red berries, confusingly, are described as both edible and poisonous, according to which guidebook one consults. The plants grow at subalpine elevations in our mountains, usually in the deep shade of evergreen forests. Twisted-stalks are found in most of the western states and across the northern part of the continent.

Western stenanthium
Stenanthium occidentale A. Gray (left)

The western stenanthium grows along streambanks and in moist meadows. It is the only *Stenanthium* native to the West (another species is found in the eastern United States). Ours occurs in northern California, British Columbia, Alberta and the four northwestern states; we have seen it growing only in the northern part of Idaho. Dainty, nodding, bronze-colored, lily-like flowers with six recurved tepals are borne on downward-curving stemlets that come off at intervals along a slender stem. Several moderately wide leaves surround the base. *Stenanthium* was derived from two Greek words meaning "narrow flower," apparently for their small size. The common name "western featherbells" has been suggested for the plant.

Western trillium (below)
Trillium ovatum Pursh

The western trillium is a spring flower that often blooms while surrounded by banks of snow. The name *ovatum* refers to its pointed, oval leaves. Its three-petaled flowers are white, turning pink and then a light purple as they mature (below). Trillium flowers vary in size, tending to be smaller at higher elevations. We have not seen trilliums growing south of the Clearwater River drainage. Meriwether Lewis collected the western trillium on the Columbia River below today's The Dalles, on April 10, 1806. The western trillium is found in all the Pacific coastal states and British Columbia, east to Montana and—rarely—in the neighboring Rocky Mountain states. On June 15, 1806, while on the Lolo Trail, Lewis and Clark collected *Trillium petiolatum* Pursh (the purple trillium, not shown). It is an uncommon species, characterized by a reddish-purple, three-petaled flower and long-stemmed, round leaves. The plant is found in the northern half of Idaho and in Washington and Oregon.

California false hellebore
Veratrum californicum **Dur.**

California false hellebores (also known as corn-lilies for their large leaves and budding flowers) are usually known simply as veratrums. They appear in late spring, growing on moist ground as high as treeline. Lewis and Clark collected the plant along the Lolo Trail in northern Idaho on June 25, 1806, but as it was not flowering it could not be classified. The Greek *elleboros* and Latin *veratrum* were used in antiquity for poisonous plants. Later, veratrums were given the Latin name *Veratrum* and plants in the buttercup family became *Helleborus*. Plants in both genera are poisonous. The poisons are pharmacologically active—*Helleborus* species contain cardiac glycosides, and veratrum extract was used to treat hypertension. *Veratrum californicum* occurs in all the western mountain and coastal states. Another variety, var. *caudatum* (A. Heller) C. L. Hitchc., also grows in Idaho. The two varieties are so similar that either will be recognized immediately as a California false hellebore.

Bear grass
Xerophyllum tenax
(Pursh) Nutt.

Bear grass (also known as Indian basket grass) is one of the Northwest's more spectacular plants. A basal tuft of leaves (the "grass") forms during snowmelt. A few weeks later, a tall stalk tipped with myriad little lily-like blooms appears. The leaves are ideally suited for basket making, for they are strong, hard and even in width almost to their tips. Lewis and Clark saw baskets woven from the grass during the winter of 1805–1806 while at the expedition's Fort Clatsop on the Columbia estuary. Then the following spring, while in today's Idaho, they saw the plants blooming. Lewis collected two intact blooms during the crossing of the Lolo Trail on June 15, 1806. Bear grass is found in all the northwestern states and Canadian provinces. We have not seen it in Idaho's mountains south of the southern approach to Lost Trail Pass (US Highway 93) on the Idaho-Montana border. Farther west it may be found in mountains near McCall, Idaho.

Elegant camus
Anticlea elegans (Pursh) Rydb.

The elegant camas (also known as mountain death camas, formerly classified as *Zigadenus elegans*) grows, sometimes in great numbers, in mountain meadows as high as treeline. Usually about a foot high, the plants may grow to twice that height in favorable situations. They flower from early to midsummer according to elevation. As the name *elegans* suggests, it is the most attractive of our three death camases. Its white flowers are large and the tepals are marked with heart-shaped green glands at their bases. All the death camases are poisonous; this one is said to be least so—also, since it grows at higher elevations it is less accessible to grazing stock. The elegant camas was collected by Meriwether Lewis in the vicinity of today's Lewis and Clark Pass near Lincoln, Montana, on July 7, 1806. Elegant camas is widely distributed, from Alaska south to Mexico in the West, east in Canada to Hudson's Bay and in the United States to the Great Lakes, with scattered populations even farther to the east.

Foothill death camas
Toxicoscordion paniculatum (Nutt.) Rydb.

The foothill death camas appears in early spring and blooms at about the same time as the common camas, usually in early June. A naked stem is topped by a cluster of small, white, six-tepaled flowers. The flowers have six anthers that protrude beyond the tepals, and three styles. Several flowers are borne on each stemlet (botanically, these are panicles, explaining its scientific species name).

The closely related and very similar meadow death camas, *Toxicoscordion venenosum* (S. Watson) Rydb.—the Latin species name means "very poisonous"—has larger, tight-clustered flowers, one flower to each stemlet. Both plants are poisonous and may be lethal to browsing animals, or to humans if they eat the roots. Native Americans were aware of this and harvested the roots of the edible common camas while both plants were in bloom to avoid potentially fatal mistakes—the roots of the common camas are quite similar to those of the death camases. Until recently the death camases were all classified as species of *Zigadenus* and are still so listed in most guidebooks. The generic name, *Toxicoscordion*, from the Greek, means "poisonous garlic." The plants are found in parts of all the Rocky Mountain states and west to the Pacific coastal states.

Orchid Family (Orchidaceae)

The word "orchid" is derived from the Latin word *orchis,* a term used by Pliny the Elder (AD 23-79) in his encyclopedia, *Natural History,* and ultimately from a similar Greek word meaning "testicle," from the shape of the pseudobulb found in many orchid plants. The family consists of 790 genera and some 18,500 species. Given such numbers, it is surprising that the Orchid family's only food use is the flavoring derived from Mexican *Vanilla* species. The family's economic importance is derived almost completely from cultivation of its spectacular flowers. Some of its members have made the step from the tropics and subtropics to colder environments while evolving progressively less showy flowers.

Orchids grow as far north as Siberia, Iceland, Greenland and throughout North America. Many of the tropical orchids are epiphytic (i.e., live on other plants), whereas those that live farther north are terrestrial. Orchids are usually perennials that spread by underground stems (rhizomes), as well as by tiny wind-sown seeds—their seeds are the smallest found in any plant. Species belonging to nine genera occur in Idaho; a few are quite showy. Our wild orchids should never be picked. Their growth requirements are highly selective and often depend on the presence of specific fungal symbionts; they will not survive transplantation.

Fairy slipper
***Calypso bulbosa* (L.) Oakes
var. *occidentalis* (Holz.) B. Boivin**

The fairy slipper is the most colorful of our orchids. It is not rare, but it is elusive. Look for it in the late spring, north of the Clearwater River drainage in Idaho. Lewis and Clark—most likely Meriwether Lewis—discovered this then-new-to-science plant blooming along the Lolo Trail and collected one as a specimen on June 16, 1806. Our plant—the image shown here—was photographed in the DeVoto Grove, west of Lolo Pass, close to where the explorers found theirs. Another variety, var. *americana* (R. Br.) Luer, occurs (rarely) in Idaho.

235

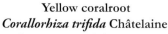

Yellow coralroot
Corallorhiza trifida Châtelaine

Because the coralroots—named for the bright red color of their root systems—are saprophytes, they lack chlorophyll. Their "leaves" are represented by sheathing bracts seen on the lower part of the stems. The color of coralroots ranges from the reddish-brown of the spotted coralroot (shown below) to the yellowish color of this plant. This plant's species name, *trifida*, means three-lobed, describing the irregular end of the flower's lip, as shown in the inset. This plant is widely distributed all across northern North America to Alaska, the arctic islands and Greenland.

Spotted coralroot
Corallorhiza maculata (Raf.) Raf.
var. *occidentalis* (Lindl.) Ames

The flowers shown here—magnified about four times lifesize—are those of a relatively common orchid that grows in moist to fairly dry, open shade of forests, where its slender, reddish-brown stems stand out. The flowers with their spotted white lips are striking, but you will need a hand lens to see them well, true of many of our native orchids. Another variety, var. *maculata,* also occurs in Idaho. It is a similar plant, with a narrower flower lip. The spotted coralroot is widely distributed across the United States and Canada, occuring in all but a few midwestern and southern states. Look for it as a brownish-red stalk standing about 12 inches high in the open shade of evergreen forests.

Alaska rein orchid
Piperia unalascensis (Spreng.) Rydb.

The Alaska rein orchid is one of several related plants character-ized by tall spike-like stems and many (usually) white blooms given off at regular intervals. We will follow the classification used by the USDA for these plants. It is a tall, stout-stemmed plant that favors moist terrain, along streams and on swampy ground. The flowers, like others in this group of plants, are resupinate (see the rattlesnake plantain below). The plant has an interesting distribution as it occurs throughout the West, from Alaska to California, and east to New Mexico (where it is rare) and South Dakota. There are also disjunct populations in Quebec, Newfoundland and Michigan, where it is rare.

Western rattlesnake plantain (left and below)
Goodyera oblongifolia Raf.

The rattlesnake plantain catches the eye only because its several deep green basal leaves are mottled with a central white stripe; it is otherwise a rather plain little plant. Apparently the name "rattlesnake plantain" comes from a fancied resem-blance of the leaves' markings (below) to those of rattlesnakes. Whitish flowers (left) form a loose cluster at the top of the plant's stem; then later in the summer one sees only the basal leaves and a dried stem. Interestingly, because the pedicels (stemlets) are twisted, the plant's flowers are upside down (resupinate) and a se-

pal is fused with two petals to form a "hood." Look for this little orchid in shaded woods. The generic name, *Goodyera,* honors English botanist, John Goodyer (1592–1664). This orchid is widely distributed in our West, north to Alaska and with a separate population in eastern Canada.

White bog orchid (left and right)
Platanthera dilatata
(Pursh) Lindl. ex L. C. Beck

The white bog orchid has several names: rein orchid, (for the strap-like lower lip), boreal orchid, and habenaria (a former generic name). The plant shown here is the most common of several varieties, var. *dilatata*. It grows from Alaska, through the northwestern states and south to Utah and Colorado. Other varieties are found throughout Canada and in the northeastern United States. Small white flowers are only one-quarter inch in diameter, but with magnification one can see that this is unmistakably an orchid. A ventral lip, two sepals extending laterally and a "hood" made up of a dorsal sepal and petals are family characteristics. A prominent downward-curving posterior spur confirms the identification. *Platanthera*, from the Greek, means "wide (or flat) anther"; *dilatata*, from the Latin, means "broad" or "wide," referring to the ventral lip's wide base.

Slender bog orchid
(left and right)
Platanthera stricta Lindl.

The slender bog orchid (formerly *Habenaria saccata* Greene), while not uncommon, is less often encountered than the plant shown above. As its former classification *saccata* suggests, it is characterized by a swollen, sac-like "scrotiform" spur that hangs down behind the flower. The species name, *stricta*, means "straight." Lindley, who described the plant, apparently used the name to refer to the narrow flower cluster. The plant occurs throughout the West, from Alaska to New Mexico.

Hooded ladies'-tresses
Spiranthes romanzoffiana Cham.

Hooded ladies'-tresses are found all across the North American continent and in most of the northern and western states. (Disjunct populations are even found in the British Isles and elsewhere in Eurasia.) In our area the plant grows as high as the alpine zone. Its stemless flowers are borne on a stout, spike-like stem. Each flower is offset a few degrees from the one below, forming the spiral that gives the plant its generic name. The petals (except for the ventral lip) join to form a "hood." *Spiranthes* was derived from two Greek words, *speira* meaning "coil" or "spiral" and *anthos* for "flower." The species name honors Nicolai von Romanzov (1734–1826), the Grand Chancellor of the Russian Empire, who sponsored—and funded—the around-the-world expedition led by explorer Otto von Kotzebue (1787–1846) in the years 1815–1818.

James L. Reveal

Mountain lady's slipper
Cypripedium montanum Douglas ex Lindl.

The mountain lady's slipper bears flowers that are smaller and more numerous than those of other species of *Cypripedium*; still, with its showy white blooms it, too, is an elegant plant. Two other lady's slippers are native to Idaho. The one shown here is the most common of the three—finding one will make a hiker's day. Meriwether Lewis encountered the mountain lady's slipper on the Lolo Trail and again at Traveler's Rest, near today's Missoula, Montana, on July 1, 1806. We do not know whether he gathered a specimen; if so, it did not survive the homeward journey. *Cypripedium montanum* grows in the mountains of Washington, Idaho, Montana and British Columbia and, rarely, in Alaska, California, Colorado and Oregon.

Miscellaneous Plants

A Selection of Miscellaneous Wildflowers

We have included here, at the end of the book, a few more wildflowers and plants that are unusual for one reason or another. Several are not well known, others do not fit into one's usual perception of what a "wildflower" should be. Nature does not always allow for our expectations!

Arctic Willow
Salix arctica Pallas

In a few of our northern mountain states, it is not unusual to find arctic plants growing on alpine tundra well above treeline. A good example is the plant that forms the white patches seen here beside a small lake in Idaho's Hyndman Basin, located in the central part of the state. The patches are neither snow nor patchy frost but flowering arctic willow plants, *Salix arctica*. The arctic willow is found in the mountains of Washington, Montana, Oregon and Idaho. Those shown here are about as far south as the plant grows. It is widely distributed to the north, however, found in most Canadian provinces and on all the arctic islands. Technically the arctic willow is a subshrub, one that flourishes close to or actually on the ground, where it forms a tangle of woody stems that bear typical willowy leaves, mixed in among the sedges that are the principal ground cover.

Tinker's penny

Hypericum anagalloides Cham. & Schltdl.

Tinker's penny, or creeping bogwort, closely resembles St. John's wort except that it is a tiny plant. Both are in the same family (Clusiaceae), but this little plant, unlike St. John's wort, is a native that has evolved as a semi-aquatic plant growing on wet, boggy ground. It is found in all the western coastal states and Canadian provinces, and in the inland states of Idaho, Montana, Nevada and New Mexico, where it grows at least as high as the sub-alpine zone. Common St. John's wort, *Hypericum perforatum* L., is also found wild in Idaho—look for it around the Salmon River hot springs near Sunbeam Dam—even though it is a non-native plant.

English plantain *Plantago lanceolata* L.

As its common name suggests, the English plantain is a nonnative plant. It is an interesting one, however, that is found throughout the United States and Canada as well as in Greenland and Eurasia. It grows at all elevations, at least as high as the montane zone in our mountains. While now considered a weed and so classified in several states, plantain has a long history of being a useful plant. Its seeds were used as a grain and its young leaves were eaten as a green. The leaves are also said to be an effective burn dressing. The common plantain is also useful as a browse plant. Animals feed on it, and it helps with erosion control. It belongs to its own plant family, Plantaginaceae.

Broadleaf cattail *Typha latifolia* L.

The broadleaf cattail's leaves are broad (*latifolia*) only in comparison with the very similar but nonnative narrowleaf cattail (*Typha angustifolia*), whose leaves are almost grass-like. The male and female parts are contiguous in the broadleaf plant, a distinguishing feature. This is the plant one will encounter in Idaho, for the narrow-leaf species occurs here only rarely. Both are otherwise widely distributed throughout the United States and Canada. They are obligate water plants, restricted to shallow, slowly moving water at the edges of streams and ponds. While various parts are edible at times, it is inadvisable to use them for food, for *Typha* are efficient collectors of toxic substances that are then stored in the plant. Because of this, they are valuable in natural water purification.

Buffalo berry
Shepherdia canadensis (L.) Nutt.

The buffalo berry (also known as russet buffalo berry, soapberry, the Indian name of soopalallie and other common names) is a relatively common bush or small tree, a member of the Elaeagnaceae, or Oleaster, family. Often encountered in our mountains, this shepherdia is characterized by its rough bark, leathery leaves and, when fruiting, by its red berries. The leaves are covered with minute scales that make them appear to have many tiny white spots, which help to identify the plants. They favor, but are not restricted to the shade of open forests. The plant is interesting chiefly because of its (barely) edible berries. These contain saponin, a soapy substance, and they are intensely bitter. Nevertheless, the berries were collected by Native Americans and mixed with other, sweeter fruit to make them palatable. They were used in this fashion either dried or fresh. When fresh, the mixture was mashed into a foamy concoction known as "Indian ice cream," which supposedly had a pleasant taste and many beneficial properties. *Shepherdia canadensis* is widely distributed, found in all the western states, across the United States in the northern tier states and in all the Canadian provinces.

Common sundew
Drosera rotundifolia L.

This image of a tangle of leaves gives little idea of what a single sundew plant looks like. Typically, a stand-alone plant consists of a rosette of leaves only a couple of inches across. A relatively tall stem with a single white flower appears as the plant matures. None of this is apparent, unfortunately, in our image. What we do see are typical sundew leaves borne on stout stemlets. The leaves are covered with hairs, each of which produces a droplet of glistening, mucilagenous fluid that attracts insects. The fluid is extremely sticky, and insects attracted by the reddish leaves and an apparent food source are caught and held in place. As they liquify, their remains provide nutritional substances that are otherwise missing in the substrate of barren boggy ground on which the plants grow. Sundews are circumboreal plants found around the northern circumference of the globe. They grow in a scattering of states across the United States—common in some, rare in others. They are also found in all the Canadian provinces, Greenland and in Eurasia. Members of a fascinating group of plant carnivores, they belong to their own family, Droseraceae.

Bibliography

Craighead, J. J., F. C. Craighead Jr. with R. J. Davis
A Field Guide to Rocky Mountain Wildflowers
Boston, Houghton Mifflin Co., 1963

Coulter, J. M.
Manual of the Botany of the Rocky Mountain Region
New York, American Book Co., 1885

Davis, R. J.
Flora of Idaho
Dubuque, IA, Wm. C. Brown Co. 1952, reprinted 1955

Duff, J. F., and R. K. Moseley
Alpine Wildflowers of the Rocky Mountains
Missoula, MT, Mountain Press Publishing Company, 1989

Earle, A. S., and J. L. Reveal
Lewis and Clark's Green World: The Expedition and Its Plants
Helena, MT, Farcountry Press, 2003

Flora of North America Editorial Committee
Flora of North America North of Mexico (30 vols.)
Vols. 1, 3–5, 19–23
New York, Oxford University Press, 1993

Gledhill, D.
The Names of Plants (3d Ed)
Cambridge, UK, Cambridge University Press, 2002

Harrington, H. D.
Edible Native Plants of the Rocky Mountains
Albuquerque, NM, University of New Mexico Press, 1967

Hitchcock, C. L., and A. Cronquist
Flora of the Pacific Northwest, An Illustrated Manual
Seattle, WA, University of Washington Press, 1973

Hitchcock, C. L., et al.
Vascular Plants of the Pacific Northwest
Seattle, WA, University of Washington Press, 5 vols. 1955,
7th printing 1994 with corrections.

Kershaw, L., A. McKinnon and J. Pojar
Plants of the Rocky Mountains
Edmonton, AB, Lone Pine Publishing, 1998

Mabberly, D. J.
The Plant Book: A Portable Dictionary of Vascular Plants (2d ed.)
Cambridge, UK, Cambridge University Press 1997
reprinted with corrections 1998, 2000

Niehaus, T. F., and C. L. Ripper
Peterson Field Guide to Pacific States Wildflowers (2d ed.)
Boston, Houghton Mifflin Harcourt Co., 1998

Philips, H. W.
Central Rocky Mountain Wildflowers and
Northern Rocky Mountain Wildflowers
Helena, MT, Falcon Publishing Company, 1999, 2001

Schreier, C.
A Field Guide to Wildflowers of the Rocky Mountains
Moose, WY, Homestead Publishing, 1996

Strickler, D.
Forest Wildflowers,
Alpine Wildflowers,
Wayside Wildflowers of the Pacific Northwest
Columbia Falls, MT, Flower Press, 1988, 1990, 1993

Strickler, D.
Northwest Penstemons
Columbia Falls, MT, Flower Press, 1997

Reveal, J. L., and J. S. Pringle.
Taxonomic Botany and Floristics
Flora of North America Editorial Committee (ed.), in *Flora of North America North of Mexico* Vol. 1, pp. 157–192
New York, Oxford University Press, 1993

Whitson, T. D., et al.
Weeds of the West (5th ed.)
Jackson Hole, WY, University of Wyoming, 1996

Zwinger, A. H., and B. E. Willard
Land Above the Trees: A Guide to American Alpine Tundra
New York, Harper & Row, 1972

Oxford English Dictionary (2d ed. on CDROM version 3.0)
New York, Oxford University Press, 2002

Useful Websites

California Plant Names
www.calflora.net/botanicalnames/

Dictionary of Botanical Epithets
www.winternet.com/~chuckg/dictionary.html
usda/fnach7.html

USDA Plants Database http://plants.usda.gov/

Index*

Achillea millefolium 58
Aconitum columbianum 162
Acteae rubra 161
Agastache urticifolia 129
Agavaceae 219
Agave family 219
Ageratina occidentale 55
Agoseris aurantiaca 33
   glauca 33
alfalfa 111
Alliaceae 219
Allium aasea 219
   acuminatum 220
   brandegeei 220
   brevistylum 220
   geyeri 221
   schoenoprasum 221
   simillimum 222
   tolmiei 222
alp-lily, common 230
alumroot, gooseberry-leaved 191
   poker 191
Amelanchier alnifolia 175
Amsinckia menziesii 67
Anaphalis margaritacea 54
androsace, Rocky Mtn 159
Androsace montana 159
anemone, cliff 163
   Piper's 164
   small-flowered 164
Anemone multifida 163
   occidentalis 163
   parviflora 164
   patens 163
   piperi 164
Angelica arguta 9
Antennaria media 54
   racemosa 55
   rosea 54
   umbrinella 54
Anticlea elegans 234
Apiaceae 9
Apocynaceae 17
Apocynum androsaemifolium 19
Aquilegia caerulea 164
   flavescens 165
   formosa 165
Arabis microphylla 71
Arctium lappa 55
   minus 55
Arctostaphylos uva-ursi 97
Arenaria kingii 87
arnica, heartleaf 30
   shining 31
   slender 31
   spearleaf 32

streambank 30
twin 31
artic willow 240
Arnica amplexicaulis 30
   cordifolia 30
   fulgens 31
   gracilis 31
   lanceolata 30
   latifolia 31
   longifolia 32
   sororia 31
Artemisia cana 59
   ludoviciana 59
   tridentata 59
Asarum caudatum 99
Asclepiadaceae 17
Asclepias fascicularis 17
   speciosa 18
Asperugo procumbens 68
Aster family 20
Aster integrifolius 22
   occidentalis 22
   scopulorum 24
aster, alpine 20, 24
   elegant 23
   hoary 23
   leafy 21
   Rocky Mountain 24
   thickstem 22
   western American 22
   white prairie 21
Asteraceae (Aster family) 20
Astragalus agrestis 104
   alpinus 104
   australis 103
   canadensis 103
   purshii 103
   vexilliflexus 104
avens, Ross's 180
bachelor's button 57
Balsamorhiza hookeri 34
   macrophylla 34
   sagittata 34
balsamroot, arrowleaf 34
   Hooker's 34
   largeleaf 34
baneberry, red 161
Barberry family 60
bear grass 232
bedstraw, fragrant 187
   intermountain 187
   northern 186
   Watson's 187
beeplant, Rocky Mt. 90
   yellow 90
Berberidaceae 60
Berberis aquifolia 60

nervosa 60
repens 60
bird's beak, bushy 200
   Yakima 200
biscuit-root, nine-leaf 12
   bare-stemmed 12
bistort, alpine 152
   American 153
Bistorta bistortoides 153
bitterbrush 177
bitterroot, alpine 156
   common 155
   pygmy 156
bladderpod, alpine 72
   western 72
blazing star 131
   Nevada 132
   ten-petaled 131
   whitestem 132
Blazing star family 131
Bleeding heart family 112
Bloomer, Hiram Green 35
bluebell, alpine 63
   ciliate 64
   Idaho 65
   leafy 66
   Oregon 66
blue-eyed grass, Idaho 218
blue-eyed Mary 204
Boechera cusickii 69
   holboellii 71
   nuttallii 71
   puberula 70
   sparsiflora 70
   williamsii 71
bog orchid, slender 238
   white 238
boneset, western 55
Borage family 61
Boraginaceae 61
Brassicaceae 69
brickellia, large-flowered 52
Brickellia grandiflora 52
brooklime, American 208
Broomrape family 194
broomrape, western naked 194
Buckthorn family 173
Buckwheat family 150
buffalo berry 242
bugbane, false 172
bugloss, viper's 68
Buglossoides arvense 68
bunchberry 91
Bunchflower family 219
burdock, greater 55
   lesser 55
burnet, salad 185

Butcher's broom family 219
butter-and-egg plant 210
buttercup, Blue Mountain 168
   graceful 171
   hillside 170
   pink 169
   plantain-leaved 169
   sage 170
   sharpleaf 171
   subalpine 170
   white water 169
Buttercup family 161
Cabbage family 69
Cactaceae 78
cactus, brittle 79
   nipple 79
   spiny prickly pear 78
Cactus family 78
Calochortaceae 219
Calochortus apiculatus 226
   elegans 226
   eurycarpus 226
   macrocarpus 227
   nuttallii
Caltha leptosepala 162
Calypso bulbosa 235
camas, common 223
Camassia quamash 223
Camissonia subacaulis 141
   tanacetifolia 141
campion, bladder 88
candyflower, Chamisso's 157
candytuft, false 74
   Idaho 74
Caprifoliaceae 80
Cardaria draba 77
Caryophyllaceae 84
cascara 173
Cassiope mertensiana 98
Castilleja angustifolia 197
   applegatei 196
   covilleana 196
   cusickii 197
   flava 195
   miniata 195
   rhexifolia 196
Cattail, broadleaf 241
   narrowleaf 241
Ceanothus sanguineus 173
   velutinus 173
Centaurea cyanus 57
   solstitialis 57
   stoebe 57
Cerastium arvense 85
   nutans 85
chaenactis, alpine 53
   Evermann's 53

Chaenactis douglasii 53
   evermannii 53
Chamerion angustifolium 138
   latifolium 138
checkermallow, Oregon 134
Chenopodiaceae 89
Chenopodium capitatum 89
chickweed, field 84, 85
   sticky 84
Chimaphila menziesii 100
   umbellata 100
Chionophila tweedyi 204
chive 221
chokecherry, western 176
Chorispora tenella 77
Chrysanthemum spp. 25
Chrysothamnus nauseosus 39
   viscidiflorus 40
Cicuta douglasii 13
cinquefoil, biennial 183
   early 182
   mountain meadow 182
   sheep 183
   shrubby 174
   slender 183
   sticky 182
Cirsium arvense 57
   foliosum 56
   hookerianum 56
   inamoenum 56
   subniveum 56
Clarkia pulchella 139
Claytonia cordifolia 157
   lanceolata 156
   parviflora 157
   perfoliata 157
   sibirica 157
clematis, hairy 166
   western 166
Clematis hirsutissima 166
   ligusticifolia 166
   occidentalis 166
Cleomaceae 90
Cleome family 90
Cleome lutea 90
   serrulata 90
clintonia 228
Clintonia uniflora 228
Clover
   long-stalked 109
   Owyhee 109
   red 111
   white 110
Cluster lily family 219, 225
Clusiaceae 241
Collinsia grandiflora 204
   parviflora 204

* If one excludes trees, grasses and non-vascular plants, the Lewis and Clark Expedition collected, or mentioned in their journals, eighty-seven flowering plants that grow in Idaho. The Lewis and Clark Expedition and Frederick Pursh, who classified the expedition's plants, are mentioned so many times in the present book that an inclusive listing becomes so long as to be self-defeating. Readers with a special interest in the expedition should consult *Lewis and Clark's Green World: The Expedition and Its Plants* by A. Scott Earle and James L. Reveal, Farcountry Press, 2003. That book treats all the Idaho plants that the explorers collected or mentioned in their journals, in detail.

Collomia, diffuse 146
  narrow-leaf 146
  staining 147
Collomia linearis 146
  tenella 146
  tinctoria 147
columbine, Colorado blue 164
  Sitka 165
  yellow 165
Compositae 20
Composite family 20
Conium maculatum 13
Corallorhiza maculata 236
  trifida 236
coralroot, spotted 236
  yellow 236
Cordylanthus capitatus 200
  ramosus 200
Cornaceae 91
cornflower 57
Cornus canadensis 91
  nuttallii 92
  sericea 92
corydalis, golden 113
Corydalis aurea 113
  caseana 112
cow parsnip 13
crane's bill 118
Crassulaceae 94
Crataegus douglasii 175
crazyweed, Bessey's 102
  late yellow (mountain) 105
  northern yellow 105
Crepis modocensis 41
cryptantha, Waterton Lakes 62
Cryptantha sobolifera 62
Currant family 119
currant, golden 120
  Henderson's 119
  Hudson's Bay 121
  sticky 121
Cymopterus glaucus 10
  nivalis 10
  douglassii 11
Cynoglossum officinale 68
Cypripedium montanum 239
daisy (see fleabane)
  Coulter's 29
  cutleaf 28
  dwarf mountain 28
  line-leaf 35
  scabland 35
  showy 26
dandelion, common 40
  false mountain 33
  orange mountain 33
Dasiphora fruticosa 174
death camas, elegant 234
  foothill 234
  meadow 234
Delphinium depauperatum 167
  glaucum 168
  nuttallianum 167
  occidentale 168
Descurainia pinnata 75
  sophia 75
desert-parsley, fernleaf 11
Dianthus caryophyllus 88

deltoides 88
Dicentra formosa 113
  uniflora 113
Disporum trachycarpum 228
dock, willow 154
Dodecatheon jeffreyi 158
  pulchellum 159
Dogbane family 17
dogbane, spreading 19
Dogwood family 91
dogwood, red osier 92
  western flowering 92
Douglasia montana 159
Draba, few-seeded 73
  globe-fruited 73
  lance-leaf 73
  Payson's 73
  Stanley Creek 73
Draba breweri 74
  cana 74
  lanceolata 73
  oligosperma 73
  paysonii 73
  sphaerocarpa 73
  trichocarpa 73
  verna 77
Dracocephalum parviflorum 129
dragonhead, American 129
Drosera rotundifolia 242
Droseraceae 242
drymocallis, cliff 182
Drymocallis pseudorupestris 182
dusty maiden 53
elder, western black 83
  western blue 83
Elaeeagnacea 242
elkweed, clustered 116
Epilobium alpinum 140
  angustifolium 138
  brachycarpum 140
  ciliatum 140
  lactiflorum 140
  obcordatum 139
  paniculatum 140
Eremogone kingii 87
Ericaceae 95
Ericameria discoidea 38
  nauseosa 39
  suffruticosa 38
Erigeron spp. 25
  asperugineus 26
  bloomeri 35
  compositus 28
  corymbosus 25
  coulteri 29
  evermannii 28
  leiomerus 25
  linearis 35
  speciosus 26
  ursinus 27
Eriogonum caespitosum 152
  capistratum 151
  flavum 151
  heracleoides 151
  ovalifolium 151
  pyrofolium 151
  umbellatum 150
  vimineum 152

eriophyllum, common 41
Eriophyllum lanatum 41
Eritrichium nanum 63
Erodium cicutarium 118
Erysimum asperum 75
  capitatum 75
Erythronium grandiflorum 225
Escobaria missouriensis 79
Eucephalus perelegans 23
European hemlock 13
Eurybia integrifolia 22
Evening primrose family 137
evening primrose, common 142
  pale 141
Fabaceae 102
fairy bells 228
fairy slipper 238
false hellebore, California 232
false solomon's seal, feathery 224
  starry 224
fiddleneck, common 67
Figwort family 194
fireweed 138
firewheel 37
fitweed, Case's 112
  Cusick's 112
Flax family 130
flax, cultivated 130
  wild blue 130
fleabane (see also daisy) 25
  Bear River 27
  cutleaf 28
  dwarf mountain 28
  Evermann's 28
  Idaho 26
  long-leaf 25
  rockslide 25
  scabland 35
foamflower, threeleaf 193
forget-me-not, arctic alpine 63
  Asian 63
  meadow 61
Fragaria virginiana 185
Frangula purshiana 173
Frasera fastigiata 116
  montana 116
  speciosa 116
Fritillaria affinis 229
  atropurpurea 229
  pudica 229
Fumariaceae 112
Fumitory family 112
gaillardia 37
Gaillardia aristata 37
  pulchella 37
Galium boreale 186
  triflorum 187
  verum 187
  watsonii 187
Gaylussacia spp. 95
Gayophytum diffusum 137
gentian, autumn dwarf 115
  explorer's 114
  one-flowered fringed 115
  pleated 115
  white 116
Gentian family 114
Gentiana affinis 115

calycosa 114
Gentianaceae 114
Gentianella amarella 115
Gentianopsis simplex 115
Geraniaceae
geranium, Richardson's 118
  sticky 118
Geranium family 118
Geranium richardsonii 118
  viscosissimum 118
Geum rossii 180
  triflorum 180
giant hyssop 129
gilia, ball-head 148
  scarlet 147
Gilia aggregata 147
globemallow, gooseberryleaf 133
  streambank 134
Glycyrrhiza glabra 15
goldenrod, mountain 42
  Rocky Mountain hairy 42
goldenweed, shrubby 38
  stemless 49
  whitestem 38
  woolly 49
Goodyera oblongifolia 237
gooseberry, Henderson's 119
  swamp black 120
Goosefoot family 89
grass-of-Parnassus, fringed 190
  palustris 190
Grindelia squarrosa 37
gromwell, field 68
Grossulariaceae 119
groundsel, ballhead 48
  dwarf arctic 45
  rocky alpine 45
  Rocky Mountain 44
  split-leaf 44
  streambank 46
groundsmoke, spreading 137
grouseberry 97
gumweed, curly cup 37
Gypsophila paniculata 8
Hackelia micrantha 61
  patens 62
Haplopappus acaulis 38
  macronema 38
  suffruticosus 38
hawksbeard, low 41
hawkweed, western 41
hawthorn, black 175
Heath family 95
heather, Merten's mountain 95
  pink mountain 95
Helianthella uniflora 36
Helianthus annuus 36
  nuttallii 36
henbane, black 212
Heracleum maximum 13
hesperochiron, dwarf 127
Hesperochiron pumilus 127
Heuchera cylindrica 191
  grossulariifolia 199
Hieracium scouleri 41
Holodiscus discolor 177

Honeysuckle family 80
honeysuckle, European 80
  trumpet 80
  Utah 81
horkelia, pinewoods 183
Horkelia fusca 183
horsemint, western 129
hound's tongue 68
hulsea, alpine 44
Hulsea algida 44
  nana 44
Hydrangea family 122
Hydrangeaceae 122
Hydrophyllaceae 123
Hydrophyllum capitatum 127
hymenoxys, stemless 43
  tundra 43
Hymenoxys acaulis 43
  grandiflora 43
Hypericum anagalloides 241
Hypericum perforatum 241
Hysocyamus niger 21
Iliamna rivularis 134
Indian blanketflower 37
Indian hemp
Indian paintbrush (see paintbrush)
Indian pipe 101
Indian potato 14
Ionactis alpina 24
  stenomeres 24
Ipomopsis aggregata 147
  congesta 148
Iridaceae 217
Iris family 217
iris, harlequin 217
  western 217
  yellow flag 217
Iris missouriensis 217
  pseudoacora 217
  versicolor 217
ivesia, Gordon's 181
Ivesia gordonii 181
  tweedyi 181
Jacob's ladder, sticky 149
  western 149
Kalmia microphylla 98
kelseya 181
Kelseya uniflora 181
kinnikinnick 97
kittentoes, mountain 209
knapweed, spotted 57
knotweed, poke 153
Lactuca pulchella 51
  sativa 51
  serriola 51
ladies'-tresses, hooded 239
lady's slipper, mountain 239
Lamiaceae 128
larkspur, slim 167
  tower 167
  upland 167
Lathyrus lanszwertii 105
laurel, mountain 98
Ledum glandulosum 96
Leguminosae 102
leptosiphon, Nuttall's 148
Leptosiphon nuttallii 148

Lesquerella alpina 72
  occidentalis 72
lettuce, blue 51
  prickly 51
Lewisia pygmaea 156
  rediviva 155
licorice root 14
Ligusticum filicinum 14
  officinale 14
  tenuifolium 14
Liliaceae 219
Lilium columbianum 227
Lily family 219
lily, checker 229
  chocolate 229
  glacier 225
  sego 226
Limosella acaulis 200
Linaceae 130
Linanthastrum nuttallii 148
Linanthus nuttallii 148
Linaria dalmatica 210
  vulgaris 210
Linum lewisii 130
  usitatissimum 130
Lithophragma bulbifera 192
  glabrum 192
  parviflorum 192
Lithospermum arvense 68
  ruderale 67
Lloydia serotina 230
Loasaceae 131
locoweed 105
Lomatium dissectum 11
  nudicaule 12
  triternatum 12
Lonicera ciliosa 80
  involucrata 81
  utahensis 81
Lop-seed family 194
lovage, fern-leaf 14
  Scotch 14
  slender-leaved 14
lousewort, bracted 199
  elephanthead 198
  sickletop 198
lupine, longspur 108
  silky 106
  silver 107
  stemless dwarf 108
Lupinus arbustus 108
  argenteus 107
  calcaratus 108
  depressus 107
  lepidus 108
  sericeus 106
Machaeranthera canescens 23
Madder family 186
Madia glomerata 48
  gracilis 48
madwort, German 68
Mahonia repens 60
Maianthemum racemosum 224
  stellatum 224
Mallow family 133
mallow, common 135
  gooseberryleaf 133
  Oregon checker 134

Malva parviflora 135
Malva sylvestris 135
Malvaceae 133
Malva parviflora 135
Mariposa lily family 219
mariposa lily, elegant 227
  sagebrush 227
  three spot 226
  white 226
marsh marigold, white 162
Matricaria discoides 52
meadowrue, western 172
meadowsweet, shinyleaf 179
Medicago sativa 111
Melanthaceae 219
Melilotus officinale 110
Mentzelia albicaulis 132
  decapetala 131
  dispersa 132
  laevicaulis 131
menziesia, rusty 95
Menziesia ferruginea 95
Mertensia alpina 63
  bella 66
  campanulata 65
  ciliata 64
  oblongifolia 66
Microseris nigrescens 40
Microsteris gracilis 146
milk-vetch, alpine 104
  bent-flowered 104
  Canadian 103
  field 104
  Indian 103
  Pursh's 103
milkweed 17, 18
  showy 18
Mimulus guttatus 202
  lewisii 201
  moschatus 203
  nanus 203
  primuloides 203
  suksdorfii 203
  tilingii 202
miner's lettuce 157
Mint family 128
Minuartia nuttallii 87
Mitella pentandra 193
  stauropetala 193
mitrewort, five-stamen 193
  side-flowered 193
mock orange, Lewis's 122
Monesis uniflora 100
monkey-flower, dwarf purple
203
  Lewis's 201
  primrose 203
  subalpine 202
  Suksdorf's 203
  yellow 202
monkshood, western 162
Monotropa hypopithys 101
  uniflora 101
Montia chamissoi 157
monument plant 116
mountain ash 175
  Sitka 175
mountain holly 60

mudwort, Owyhee 200
mugwort, western 59
mule's ears, white 50
  yellow 50
Mulgedium oblongifolium 51
mullein, common 210
musk-flower 203
mustard, blue 77
  western tansy 75
Mustard family 69
Myosotis asiatica 63
Nasturtium officinale 76
Navarretia, Brewer's 144
  mountain 144
Navarretia breweri 144
  divaricata 144
Nicotiana attenuata 211
nightshade, climbing 212
  cutleaf 212
Nightshade family 211
ninebark, mallow-leaf 177
Noccaea fendleri 74
  montana 74
Nothocalais spp. 40
  nigrescens 40
Nuphar polysepala 136
Nymphaeaceae 136
Ocean spray, hillside 177
Oenothera caespitosa 142
  pallida 141
  subacaulis 141
  tanacetifolia 141
  villosa 142
Oleaster family 242
Onagraceae 137
onion, Aase's 219
  Brandegee's 220
  Geyer's 221
  Hooker's 220
  short-styled 220
  simil 222
  Tolmie's 222
Onion family 219
Onobrychis viciifolia 111
Opuntia fragilis 79
  polyacantha 78
Orchid family 235
orchid, slender 238
  white bog 238
Orchidaceae 235
Oregon grape, creeping 60
Oreostemma, alpigenum 20
Orobanchaceae 194
Orobanche uniflora 194
Orogenia linearifolia 14
Orthilia secunda 99
Orthocarpus tenuifolius 198
Osmorhiza berteroi 15
  occidentalis 15
owl's-clover, thin-leaf 198
Oxytropis besseyi 102
  campestris 105
  monticola 105
Oyster plant 51
Packera dimorphophylla 45
  pseudoaurea 46
  streptanthifolia 44
  subnuda 45

werneriifolia 45
Paeonia brownii 143
Paeoniaceae 143
paintbrush, Coville's 196
  Cusick's 197
  lemon yellow 197
  narrow leaf 197
  rosy 196
  scarlet 195
  wavy leaf 196
Parnassia fimbriata 190
Parsley family 9
pasqueflower, American 163
  western 163
Pea family 102
pea, Nevada 105
pearly-everlasting 54
Pedicularis bracteosa 198
  groenlandica 198
  racemosa 198
pennycress, field 77
penstemon, dark blue 206
  fuzzy-tongue 206
  hot-rock 206
  mountain 207
  pale yellow 205
  Payette 206
  shrubby 207
  small-flowered 206
  taper-leaf 205
  twin-leaved 206
  Wilcox's 206
Penstemon attenuatus 205
  confertus 205
  cyanus 206
  deustus 206
  diphyllus 206
  eriantherus 206
  fruticosus 207
  montanus 207
  payettensis 206
  procerus 206
  wilcoxii 206
Peony family 143
peony, western 143
Perideridia montana 16
Persicaria vivipara 152
phacelia, Franklin's 126
  Idaho 126
  silverleaf 124
  thread-leaf 126
Phacelia franklinii 126
  hastata 124
  heterophylla 125
  idahoensis 126
  linearis 126
  sericea 123
Philadelphus lewisii 122
Phlox family 144
phlox, longleaf 145
  showy 145
  slender 145
  spreading 145
Phlox diffusa 145
  gracilis 145
  hoodii 145
  longifolia 145
  speciosa 145

Phrymaceae 194
Phyllodoce empetriformis 98
Physaria didymocarpa 72
  occidentalis 72
  reediana 72
Physocarpus malvaceus 177
pineapple weed 52
pinesap 101
Pink family 84
pink, maiden 88
Piperia unalascensis 237
pipsissewa 100
Plagiobothrys tenellus 62
Plantage lanceolata 241
Plantaginaceae 194, 241
plantain, English 241
Plantain family 194, 241
Platanthera dilatata 238
  stricta 238
plum, American 176
Polemoniaceae 144
polemonium, showy 149
  sticky 149
Polemonium occidentale 149
  pulcherrimum 149
  viscosum 149
Polygonaceae 150
Polygonum phytolaccifolium 153
  viviparum 152
pond-lily, Rocky Mountain 136
popcorn flower, slender 62
Portulacaceae 155
Potentilla biennis 183
  concinna 182
  diversifolia 182
  glandulosa 182
  gracilis 183
  ovina 184
prairie-dandelion, black-hairy 40
prairie smoke 180
prairiestar 192
Primrose family 158
primrose, Cusick's 160
  Parry's 160
Primula cusickiana 160
  parryi 160
Primulaceae 158
prince's pine 100
Prosartes trachycarpa 228
Prunus americana 176
  virginiana 176
Pseudostellaria jamesiana 84
puccoon, Columbia 67
purple fringe 123
Purshia tridentata 177
Purslane family 155
pussy-toes, Hooker's 54
  Rocky Mt. 54
  rosy 54
pyrethrum 52
Pyrola asarifolia 99
  chlorantha 99
  picta 99
rabbitbrush, green 40
  rubber 39
ragged robin (pink fairies) 139
ragweed, arrowhead
  tall 47

ragwort, alkali-marsh 46
 ballhead 48
 dwarf mountain 4
ranger's buttons 16
Ranunculaceae 161
Ranunculus acriformis 171
 alismifolius 169
 andersonii 169
 aquatilis 169
 eschscholtzii 170
 flammula 171
 glaberrimus 170
 inamoenus 171
 jovis 170
 oresterus 168
raspberry, common 178
rattlesnake plantain, western 237
rein orchid, Alaska 237
Rhamnaceae 173
Rhododendron californicum 95
 neoglandulosum 95, 96
Ribes aureum 120
 cereum 120
 hudsonianum 121
 lacustre 120
 oxycanthoides 119
 viscosissimum 121
rock-rose 142
rockcress, Cusick's 69
 elegant 70
 hoary 70
 Holboell's 71
 Nuttall's 71
 Williams's 71
 Wind River 71
rock-fringe 139
rocket, yellow 76
Rorippa nasturtium-aquaticum 76
Rosa nutkana 179
 woodsii 178
Rosaceae 174
Rose family 174
rose, Nootka 179
 Wood's 178
Rubiaceae 186
Rubus idaeus 178
 parviflorus 178
Rumex paucifolius 154
 salicifolius 154
Ruscaceae 219
sage, big 59
 Louisiana 59
 silver 59
sainfoin 111
St. John's wort 241
Salicaceae 241
Salix arctica 240
salsify, yellow 51
Sambucus cerulea 83
 racemosa 83
sandwort, Uinta 87
Sanguisorba minor 185
Saponaria officinalis 84, 88
Saskatoon serviceberry 175
Saxifraga integrifolia 189
 occidentalis 189
 odontoloma 189
 oppositifolia 190

rhomboidea 188
Saxifragaceae 188
Saxifrage family 188
saxifrage, brook 189
 Columbian 189
 diamondleaf 188
 mountain 189
 purple 190
 wholeleaf 189
scorpion weed, variable-leaf 125
Scrophulariaceae 194
Scutellaria angustifolia 128
 antirrhinoides 128
Sedum lanceolatum 94
 stenopetalum 94
Senecio spp. 44
 cymbalaria 44
 fremontii 47
 hydrophilus 46
 serra 47
 sphaerocephalus 48
 triangularis 47
serviceberry, western 175
Shepherdia canadensis 242
shepherd's purse 77
shooting star, Jeffrey's 158
 many-flowered 159
Sibbald, Robert 184
Sibbaldia procumbens 184
Sidalcea oregona 134
Silene, Menzies's 87
 Oregon 86
 Parry's 86
 Scouler's 86
Silene menziesii 87
 oregana 86
 parryi 86
 scouleri 86
 vulgaris 88
single-delight
Sisyrinchium idahoensis 218
 occidentale 218
skullcap, narrow-leaved 128
snapdragon 128
smelowskia, alpine 74
Smelowskia americana 74
 calycina 74
snakeroot, western 55
Snapdragon family 194
snowberry, common 82
 mountain 82
snowlover, Tweedy's 204
soap berry 242
soapwort, common 88
Solanaceae 211
Solanum dulcamara 212
 triflorum 212
Solidago lepida 42
 multiradiata 42
sopalallie 242
Sorbus scopulina 175
 sitchensis 175
sorrel, mountain sheep 154
Spiraea, betulifolia 179
 splendens 179
Spiranthes romanzoffiana 239
spirea, subalpine 179
 white 179

springbeauty, heartleaf 157
 lanceleaf 156
 Siberian 157
 streambank 157
spring-parsley, Douglass's 11
 snowline 10
 waxy 10
squawberry 120
star thistle, yellow 57
starwort, long-stalk 85
steer's head 113
Stellaria jamesiana 84
 longipes 85
 media 84
stenanthium, western 231
Stenanthium occidentale 231
Stenotus acaulis 49
 lanuginosus 49
stickseed, spreading 62
Stonecrop family 94
stonecrop, lanceleaf 94
 wormleaf 94
stoneseed 76
strawberry, wild 185
strawberry blite 89
Streptopus amplexifolius 230
sugar bowl 166
suncup, northern 141
 tansy-leaved 141
sundew, common 242
sunflower, common 36
 Nuttall's 36
 Rocky Mountain dwarf 36
swamp white-heads 16
sweet cicely, mountain 15
 western 15
sweetclover, yellow 110
swertia 117
Swertia perennis 117
Symphoricarpos albus 82
 oreophilus 82
Symphyotrichum falcatum 21
 foliaceum 21
Synthyris missurica 209
syringa 122
Tanacetum spp. 25
Tanacetum vulgare 52
tansy, common 52
Taraxacum officinale 33
tarweed, mountain 48
 slender 48
Tetraneuria acaulis 43
 grandiflora 43
Thalictrum occidentale 172
Themidaceae 219
thermopsis, mountain 106
Thermopsis rhombifolia 106
thimbleberry 178
thistle, Canadian 57
 elk 56
 Hooker's 56
 Jackson Hole 56
Thlaspi arvense 77
 idahoense 74
Tiarella trifoliata 193
Tinker's penny 241
toadflax, Dalmatian 210
tobacco, coyote 211

tobacco brush 173
townsendia, mountain 29
Townsendia alpigena 29
Toxicoscordion paniculatum 234
Tragopogon dubius 51
 porrifolius 51
trapper's tea 96
Trautvetteria carolinensis 172
Trifolium longipes 109
 owyheensis 109
 pratense 111
 repens 110
trillium, purple 231
 western 231
Trillium ovatum 231
 petiolatum 231
triplet-lily, large-flowered 225
Triteleia grandiflora 225
twinberry 81
twinpod, common 72
twistedstalk, clasping-leaf 230
Typha angustifolia 241
Typha latifolia 241
Typhaceae 241
Umbelliferae 9
Urtica dioca 129
Vaccinium spp. 95
 scoparium 97
valerian, edible 213
 northern 214
 sharpleaf 214
 Sitka 214
Valerian family 213
Valeriana acutiloba 214
 dioica 214
 edulis 213
 sitchensis 214
Valerianaceae
vase flower 166
Veratrum californicum 232
Verbascum thapsus 210
Veronica americana 208
 cusickii 208
 wormskjoldii 209
vetch, common (tare) 109
 purple-flowered woolly 109
Vicia sativa 109
 villosa 109
Viola adunca 215
 glabella 216
 macloskeyi 216
 purpurea 216
 vallicola 216
Violaceae 215
violet, goosefoot yellow 216
 hooked 215
 pioneer 216
 small white 216
 valley 216
Violet family 215
viper's bugloss
virgin's bower, western 166
wallflower, Pursh's 75
 western 75
water hemlock, European 13
 western 13
Water-lily family 136
watercress 76

Waterleaf family 123
waterleaf, ballhead 127
whiteheads, swamp 16
whitetop 17
whitlow-grass, spring 77
wild buckwheat, dwarf 151
 hairy Shasta 151
 Hitchcock's 151
 matted 152
 parsnip-flower 151
 Piper's 151
 purple cushion 151
 sulphurflower 151
 wicker-stem 152
willow, arctic 240
willow-herb, red 138
willow-weed, alpine
 autumn
 common
wintercress, American 76
wintergreen, green 99
 one-flower 100
 pink 99
 sidebells 99
 white-vein 99
woodlandstar, bulbous 192
 smallflower 192
Wyethia amplexicaulis 50
 helianthoides 50
Xerophyllum tenax 233
yampah, northern 16
yarrow, common 58
yellow bell 229
Zigadenus elegans 234
 paniculatus 234
 venenosus 23

247

About the Authors

Dr. Scott Earle, a retired academic surgeon, has devoted his retirement years to photographing Idaho's wildflowers. This book, the third edition of Idaho Mountain Wildflowers: A Photographic Compendium, *contains almost 700 of his wildflower images as well as descriptions of the included plants.*

Jane Lundin, a retired school librarian, is a devoted plant-lover who has been part of this endeavor since the the first edition of Idaho Mountain Wildflowers *was published more than a decade ago. She has found and identified many of the plants included in this and previous editions and provided many of the images used to illustrate this book.*